T0297370

Picture This!

Grasping the Dimensions of Time and Space

Picture This!

Grasping the Dimensions of Time and Space

By Michael Carroll, with original illustrations by the author

 Springer

Michael Carroll
Littleton, CO, USA

ISBN 978-3-319-24905-6   ISBN 978-3-319-24907-0 (eBook)
DOI 10.1007/978-3-319-24907-0

Library of Congress Control Number: 2016936414

Cover image by the author

Printed on acid-free paper

This Springer imprint is published by Springer Nature
The registered company is Springer International Publishing AG Switzerland

Preface

THE LONG AND SHORT OF IT

As we have become aware of our place in the universe, and its true scale, we are forced to use larger and larger numbers. Our neighborhood has expanded from the homes in our village to the planets in our Solar System to the galaxies in our local group. In our modern, technological culture, we throw around some big numbers. We have to. Our complex everyday lives necessitate it. Take, for example, our transportation. Maybe you drive a mid-sized hybrid car. Its engine turns out about 90 horse-power. If we run the clock back a century to a family similar to yours, one or two horses typically served that family's transportation needs, and perhaps a few more if they used farming equipment. What could they have done with 90 horses?

If you climb aboard a mid-sized civilian airliner, like an Airbus A320, each of its engines has power equivalent to roughly 1600 cars or 16,000 horses.[1] In a single, modest commercial airliner, we have half as many horses as were employed by the entire combined armies of Genghis Khan. The numbers we use in our everyday lives have multiplied over time. We know more. We know how small we are.

This knowledge did not come easily. The Greek scholars Aristarchus and Hipparchus were among the first who attempted to figure out the distance from our own world to the Moon. Aristarchus assumed – correctly – that when the Moon was exactly half illuminated, it formed a right angle between the Sun and Earth. By figuring the angle between the Sun and Moon, he could then assume a triangle with the three cosmic objects. The measure of this triangle provided him with an estimate of the distances between Earth and the Sun and Earth and the Moon. In the second century B. C., Hipparchus used similar methods to those used by Aristarchus a century before. He agreed with Aristarchus that the Earth-Moon distance was roughly 59 times the radius of Earth,[2] and the range to the Sun at 1200 times that of Earth's radius. He was off by 23,250 times.

1. Calculated from Airbus A320's IAE V2500 engine thrust of 155 kN.

2. Hipparchus actually estimated the distance from Earth to the Moon at 67 Earth radii.

Astronomers continued to refine their observations and math skills, gathering insights into just how large an affair the vault of the heavens truly is. Two sixteenth-century astronomers, Christiaan Huygens and Jean-Dominique Cassini, independently determined the distance to the Sun from Earth. Huygens used a very clever geometric method. Like all good telescope observers of the day, he knew that Venus exhibited phases just like the Moon does. He also realized that the phase of Venus depended on the angle between it, the Sun and Earth, just as Hipparchus had done with the Moon and Earth. So, for example, when Venus, the Sun, and Earth form a right angle, Venus looks half lit, like a half Moon. By measuring two inside angles in a triangle, you can also measure the length of one side, if you know how big the other side is. Unfortunately, he had to guess at this number. He was not far off. Cassini came along and, armed with a newly calculated distance to Mars using parallax, was able to estimate the Sun's distance at 140,000 km.

Using even more math, German mathematician/astrologer Johann Kepler[3] described the laws of motion, enabling observers to estimate the distance of planets from the Sun by the way they moved in their orbits. The Solar System turned out to be a surprisingly large affair. Space beyond was even more vast (see Chap. 6).

How do we get our minds around such concepts? We find astronomical numbers baffling. The human brain can visualize a few tens of objects on a table, but beyond that point, it begins to generalize in an effort to make comprehension manageable. The stars visible in the night sky, away from light pollution, number approximately 6000 to the naked eye. But this number is a fairly abstract one, understandable on an intuitive level only as a visual experience or impression.

However, we *can* understand by analogy and comparison. When we are told that a million Earths would fit inside the Sun, we are impressed, but a bit mystified. But when we use the comparison of a basketball next to the head of a pin, we can comprehend that relationship. *Picture This!* takes us on such a journey. By using objects and landmarks familiar to us, we will delve into the sizes, shapes, and relationships of cosmic objects that boggle the mind, humble the heart, and inspire the spirit.

It is a daunting challenge. Take, for example, the Solar System. With a basketball Sun nestled in the center, the span of our planetary system – out to Pluto – stretches 24 m across. With our pinhead-sized Earth, it is a big neighborhood. But we have only begun. The scales of our cosmic surroundings become ever more difficult to comprehend. Demonstrating the distance to the nearest star is nearly impossible, even on this diminutive scale. The nearest sun resides a substantial way across the continental United States. Only a handful of stars would fit within the entire globe of Earth. To comprehend, we must "jump scales" to something larger, shrink the Sun farther down, and compress distance into something manageable. "You have to do all these little tricks," says astrophysicist/educator Jeff

3. Kepler, who lived from 1571 to 1630, played a key role in the seventeenth-century "scientific revolution." His work served as part of the foundation for Isaac Newton's laws of physics.

Bennett. "Anything you do to try to get your head around them works for only so long. For example, we'll usually start out with the Earth-Moon system. The Moon is 30 Earth diameters away, and it is about a third the size of Earth, but then when you try to put the Sun in, you can't do it. So you have to change scale. You come up with this new scale. You can jump scale and put the galaxy on a football field, for example, but it becomes meaningless to most people. The numbers become mind numbing. So instead, I like to use counting. Counting to 100 billion takes you 3000 years. That's something you can understand."

It is difficult and often confusing to communicate relative size using the written word. For example, in 2005, the small asteroid 2005YU55 passed within 201,700 miles of Earth – closer than the Moon. Astronomers imaged the cosmic stone potato with radar, and the press described the object as being "as long as an aircraft carrier." The problem with this description is that it only involves one dimension: length, while the object is a three-dimensional one. This book depends upon visual analogies, so the best way to show the size of 2005YU55 was to place the entire object, or an approximation of it, next to an aircraft carrier. First, we place a photo of an aircraft carrier at the bottom of the frame, with a wake and a sky painted in. The photo used is the Nimitz-class aircraft carrier *USS Ronald Reagan*.

Next, we place the radar image of the asteroid above the ship, with a shadowed side and color filled in.

Fig. P.1 Asteroid 2005YU55 next to an aircraft carrier (Photo by Mass Communication Specialist 2nd Class Joseph M. Buliavac, courtesy of the U. S. Navy, ID 090608-N-3659B-059. Radar image Courtesy of NASA/JPL/Caltech.)

To provide a three-dimensional relationship between the floating rock and the surface, we drop in a shadow across the ocean and the carrier. In the radar image, craters are not visible, but they are subtly implied, and bodies of this size often seem to have them, so craters are artistically draped over the sunlit surface.

To complete the effect, reflected light from the ocean is rendered on the shadowed side. Now, we have a view of asteroid 2005YU55 in a setting that is understandable on an intuitive level.

Fig. P.2   A shadow presumably from a crater and crater texture are added to make the rock more realistic (Art by the author.)

Fig. P.3   The final result (Art by the author.)

As we explore scales of various sizes, we notice echoes in patterns of varying sizes. For example, the spiral of a hurricane reminds us of the spiral arms of a galaxy's stars or the spirals found in certain biological structures. The ancient Greeks noticed these parallels millennia ago. They coined the term "golden mean" or "golden ratio" to describe a repeating proportion found in many natural forms. Greek philosophers believed that beauty contained three essential elements: symmetry, proportion, and harmony. To describe the natural proportion of forms in nature, Greek mathematicians established the golden ratio. If two objects – compared to each other – had an identical ratio to the sum of those objects compared to the larger of the two original objects, the objects were defined as being in the golden mean. Like a computer program's fractals, the shapes repeat, no matter the scale. As a modern example, in 2010, the journal *Science* reported that the golden ratio is present at the atomic scale in the magnetic resonance of spins in cobalt niobate crystals.

A more mathematically advanced approach to the spiral forms found in nature, especially those found in biological forms such as the organization of leaves in an aloe plant (called phyllotaxis), the configuration of an artichoke's plates, or the arrangement of a pinecone's petals. This approach can be summed up by a relationship called Fibonacci sequence. The first two numbers in the Fibonacci sequence are 1 and 1.[4] Each succeeding number is the sum of the previous two (so the sequence begins as 1, 1, 2, 3, 5, 8, 13, and so on). Sunflowers famously display two superimposed opposite spirals in the florets outside of the seed head, one spiral of 34 and the other of 55 flowers. The Fibonacci sequence even shows up in the family tree of honeybees. If an unfertilized female bee lays an egg, that egg will hatch a male, or drone, bee. But if the female has mated, her egg will become a female. This means that any male bee always has one parent, 2 grandparents, 3 great-grandparents, 5 great-great-grandparents, etc. Fibonacci numerical sequence presents on many varied levels, including physics, chemistry, biology, and mathematics. For example, a physicist will tell us that the spiral is the lowest energy form that arises through self-organizing processes. The biologist sees the spiral of leaves as the most efficient use of space for leaves to gather light in photosynthesis. Structure echoes purpose.

These spiral forms follow the shape they do, in part, because their structure follows their purpose. In the same way, the spiral form of a galaxy issues from its nature: density waves move throughout the stars at its disk, resulting in its grand spirals of stars. The rings of Saturn, actually a vast spiral around the planet, are also triggered by density waves.

The concept of structure echoing purpose originally came from architecture. The early twentieth-century architect Louis Sullivan championed the idea, which he put in this way: "Form follows function." In 1896, he wrote,

4.Some applications begin the sequence at 1 and 0.

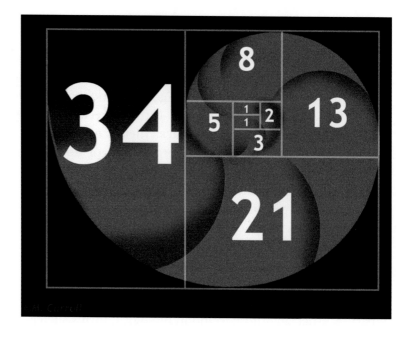

Whether it be the sweeping eagle in his flight, or the open apple-blossom, the toiling work-horse, the blithe swan, the branching oak, the winding stream at its base, the drifting clouds, over all the coursing sun, form ever follows function, and this is the law. Where function does not change, form does not change…It is the pervading law of all things organic and inorganic, of all things physical and metaphysical…that the life is recognizable in its expression, that form ever follows function. This is the law.[5]

Although the exacting numbers of Fibonacci sequence do not apply to the arms of a spiral galaxy, many natural forms bear similarity through an assortment of scale, including spirals, spheres, and disks, each form following its function or nature. We will see all of these forms repeated as we work our way through the different scales of the universe.

Just as we have come to appreciate the true size of objects in the cosmos, another daily yardstick has expanded within our life experience, that of distance. When dealing with the cosmos, distances are so great that they become linked to what is known as deep time. For example, the nearest star, Proxima Centauri,[6] is 38,900,000,000,000 km away (see Chap. 5). But a better measure of its distance is by clocking the time it takes for light to travel between Proxima and us. Light travels 299,000 km (186,000 mi.) each second, so we may also say that Proxima Centauri is 4.2 light-years away, the distance that light travels in 4.2 years.

The farther away an object is, the more ancient the light we are seeing from it. The images of Proxima Centauri in our telescopes tell us what the star was up to some 4 years ago. If we gaze at the Andromeda Galaxy, we are seeing its great star-pinwheel as it appeared 2.5 million years in the past. Likewise, anyone on a planet in the Andromeda Galaxy looking back on Earth would see things here as they were 2.5 million years ago.

5. Sullivan, Louis H. (1896). "The Tall Office Building Artistically Considered," *Lippincott's Magazine* (March 1896): 403–409.

6. Proxima Centauri is part of a three star system. It is a red dwarf orbiting around two much larger yellow stars, aCentauri A and aCentauri B. In its current orbital position, it is the closest to Earth of the three.

Fig. P.5   Similar patterns on vastly different scales. From left to right: a spiral galaxy (STScI), Hurricane Ike (NASA/JSC), and the leaves of an aloe plant (Stan Shebs, Wikipedia Commons, https://commons.wikimedia.org/wiki/File:Aloe_polyphylla_1.jpg.)

As we delve into deep time, those numbers become intertwined with our normal life experience as well. Time, in fact, is another way we can understand large numbers. For example, the number one million, by itself, is so enormous that it is a bewildering, abstract concept for us. But we can begin to get a handle on it if we convert it into a unit of time: one million seconds ago was 12 days ago. A billion seconds ago was about 18 years ago. What were you doing then? Two billion seconds ago perhaps you were graduating high school. A trillion seconds ago – 1000 billion[7] – takes us to the year 29,700 B.C. At this time, the first humans began to populate New Guinea, and we find the first evidence of domesticated dogs. Tell your dachshund. A quadrillion – 1000 billion – runs us back to 30,800,000,000 years ago, long before the universe existed.[8] These huge numbers may seem ridiculous to bandy around, and yet they continually confront us as we explore the scale of things astronomical.

Closer to home, light is too large a yardstick for measuring planetary distances, and miles and kilometers are too small. Instead, we use astronomical units. One astronomical unit (AU) is the distance between Earth and the Sun, 149 million km (which is, incidentally, a little over 8 light-minutes). So while the planet Mercury is 0.39 AU from the Sun, Neptune is a chilly 30 AU. As we speak of our Solar System and its planets and moons, we will often be using astronomical units. The planetary progression follows this pattern: Mercury orbits 0.39 AU from the Sun, Venus at 0.72 AU, Earth 1 AU, Mars 1.5 AU, Jupiter 5.2 AU, Saturn 9.6 AU, Uranus 19.2 AU, Neptune 30 AU, dwarf planet Pluto 44 AU (at its farthest), and the Kuiper Belt at a remote 50 AU at its outer edge.

7. Here, we are using what is called the short count of numbers. Some European and Asian countries use the long count, which is slightly different, but it is the same idea.

8. Current estimates put the universe at approximately 13.7 billion years old.

9. This analogy pertains to the northern hemisphere. If you live south of the equator, simply reverse the details. You face the same issue.

The comprehension of scale is critical to the understanding of the cosmos, says Jeffrey Bennett. "Everything you want people to understand about astronomy requires scale at its base. Unless you start with the scale, nothing else is going to make sense properly." As a simple example, Bennett cites a common misconception related to the seasons. Many people think Earth is closer to the Sun in the summer,[9] but seasons are actually due to Earth's tilted axis of spin. Contributing to this widespread misunderstanding is the classic textbook diagram showing the tilt of Earth's axis, with sunlight hitting the northern hemisphere more directly than the southern. "The typical reaction is, 'Well, wait a minute: the northern hemisphere *is* closer to the Sun than the southern hemisphere.' When you draw it like that, that's exactly what it looks like. In order to understand why that diagram is not the full story, you need to have done the scale, to see just how far Earth is from the Sun. When you understand [that long Earth-Sun distance], you say, 'Oh, of course it makes no difference whether you're in the northern or the southern hemisphere.' The difference is just too small. You have to show that distance before you look at that diagram or else people won't believe you."

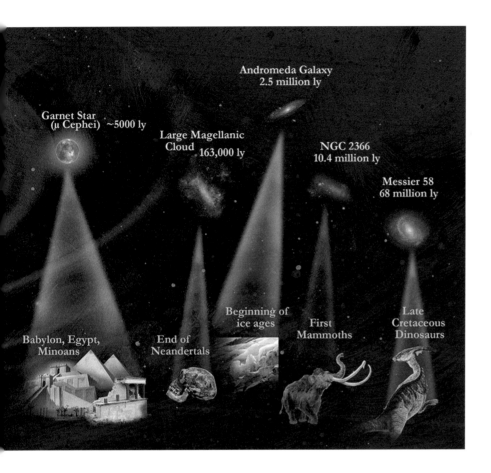

We did not always understand the universe as we do now, and in the future our twenty-first century views will appear quaint. It is always informative to look back at how our perspective of the cosmos – and our predictions of how we would explore it – has evolved over time. Our final chapters will do just that, as we examine early concepts of the worlds around us, as well as the technologies that would take us to them.

Littleton, CO, USA   Michael Carroll

Acknowledgments

After seeing some of my comparison pieces, my wife Caroline suggested the idea of a book. It was a great suggestion which I brought to my editor, Maury Solomon. Maury championed the idea at Springer, and the rest of its history resulted in the pages that follow. Ron Miller and David A. Hardy both offered valuable artistic and permissions help, and Bruno Stanek and Arthur Woods were of special assistance for getting me in touch with Olga Shonova, who generously gave permission to use her uncle-in-law Ludek Pesek's beautiful images. Likewise, Randy Liebermann played a crucial role in my obtaining the wonderful Fred Feeman image for Chap. 9. Special commendation goes to Mariecris Gatlabayan of Bonestell LLC. My thanks to Chris Calle for lending his father's masterpiece of Venus. Robert Zubrin provided me with the concept of the relationship between cars, horses, and jet planes, which also appears – in slightly different form – in my book *Living Among Giants*. Nadia Imbert-Vier played a key, generous role in helping me obtain the perfect images from the European Space Agency. Garry Hayes lent his beautiful photos of Mt St Helens, which I mercilessly chopped up for the Tycho Crater comparison. As always, Don Mitchell was there to supply a beautiful Venera image; thanks, Don! My artist/scientist friend Dr. Dirk Terrell took hold of my rudderless mathematics, making it possible for the construction of my golf ball Solar System in Chap. 5. Marilyn Flynn lent her trusty cleaver to the text, making it something far better than it would have been. Not to be outsnipped, my daughter Alexandra also pitched in on valuable editorial help. And thanks to Gordo for feline assistance.

Unless otherwise noted, all images are by the author.

Contents

Part I
Our Place in the Cosmos

Chapter 1

Asteroids, Comets and Our Cosmic Landscape

SOME SOLAR SYSTEM HISTORY

To understand the overall scale and distances of our planetary system and all its inhabitants, including the planets, asteroids and comets, we must first get a little background. The early solar nebula, the great cloud of dust and gas from which all planets and moons and other objects came, originally lacked any big planets. The Solar System began as a great cloud of interstellar gas, much like many of the beautiful nebulae we see today (see Chap. 6). Those nebulae provided clues to early theorists about how our Solar System came together. Among them, Swedish scientist Emmanuel Swedenborg proposed a simple "nebular hypothesis" in 1734. His theory described a hot globe of material around the infant Sun as the birthplace of the Solar System. German philosopher Immanuel Kant (1755) later enhanced his theories. French astronomer Pierre-Simon Laplace (1796) added even more detail, suggesting that the primordial cloud surrounding the Sun somehow flattened into a disk, eventually leading to the genesis of the planetary system.

The dynamics of a transformation from hot, spherical cloud into a flat, planet-forming disk were not well understood, and the theory had many detractors. Other theories also involved nebulae. For example, Soviet theorist Otto Schmidt suggested that the primordial Sun passed through a cosmic cloud of gas, dragging it along in a great tail that ultimately condensed into the planets we see today. In 1917, James Jeans hypothesized that a star passed close to the Sun, pulling material from it. This theory seemed to fit the outline of the planets, as if a doughnut-shaped cloud surrounding the Sun, thickening toward its outside edge, led to the small terrestrials on one side, and the large gas and ice giant planets on the far rim.

Astronomer Forest Moulton and geologist Thomas Chamberlin suggested that a passing star caused a tidal bulge in the Sun, and that this bulge streamed out into tendrils of material, which mixed with similar trails of material from the passing star. These congealed into small blobs that they called planetesimals, which eventually merged into larger planets.

Computer modeling and advances in mathematics enabled astrophysicists to settle on a Solar System formation model much like Kant's. Then, in 1993, the Hubble Space Telescope caught several young stars in action, complete with disks of condensing material just like those described in Kant's nebular hypothesis. These protoplanetary disks are also known as proplyds.

Today, we have a much better understanding of how a solar nebula becomes a system of planets. Initially eddies and currents in the interstellar cloud are triggered by the shockwaves from nearby passing – or even exploding – stars. As the waves ripple through the cloud, matter compresses.

M. Carroll, *Picture This!: Grasping the Dimensions of Time and Space*,
DOI 10.1007/978-3-319-24907-0_1, © Springer International Publishing Switzerland 2016

As our own solar nebula coalesced, the cosmic mist contracted under its own gravitation, static electrical attraction and other factors. Condensing into a more concentrated cloud, it began to spin, in the same way that an ice skater spins at an increasing rate when pulling his or her outstretched arms to the sides of the body. Like the clay on a potter's wheel, that rotation flattened the cloud into a great disk, called an accretion disk. Within that vaporous disk, some objects attracted others to themselves through gravitational or other forces, and objects began to grow. The growth became a runaway effect where the biggest object grew the fastest because it had the most mass, so it could grab other objects better than other nearby bodies. This growth continued, and the disk dissipated until there was nothing else from which to grow. Objects that were not absorbed by the cores of giant planets were deflected by those cores and flung out of the Solar System. The still-forming protostar at the center began to shrink, and in doing so developed a rapid spin. The new planets, plus the residual cloud around the center, also spinning, dragged on the star, forcing it to slow. When the pressure and temperature within the protostar rose to a sufficient level, reaching several million degrees, the atoms in the core collapsed into each other, and nuclear fusion ignited, powering the sphere into stardom.

Planetesimals going around the Sun 4.5 billion years ago had one of two likely fates: either to bump into the growing core of a giant planet and become part of it, or to be scattered by that core and ejected from the Solar System completely. Of all the objects that were kicked outward, at least 90 % left the Solar System. Today, they are off drifting among the stars as interstellar comets (the likes of which we haven't seen yet). A small fraction – not more than 10 % of those ejected bodies – were deflected a second time by nearby stars and captured into the Oort Cloud. The Oort Cloud began close in to the Sun, and then those objects were scattered out. It is likely that parts of the Kuiper Belt were scattered out in similar fashion (Fig. 1.1).

Even the planets may not have formed in the places they inhabit today. We get a hint of the wacky development of our planetary system from the Kuiper Belt itself, where many objects follow paths that are "in resonance" with Neptune. This means that for every three times Neptune circles the Sun, a more distant object "in resonance" will orbit it exactly twice. Pluto is one of the objects in resonance with Neptune. Computer modelers and planetary dynamicists realized early on that the only way to trap Kuiper Belt objects (KBOs) in resonances with Neptune would be to slowly move Neptune out from closer to the Sun.

The Kuiper Belt observations gave rise to an important new idea: the concept of planetary migrations. "Before that," says astronomer David Jewitt, "everybody was happy thinking of the Solar System as a static entity with the planets going around in the same direction and in the same orbits for all time, like a giant clock. But the high population of resonant KBOs could only reasonably be explained by moving Neptune outwards slowly

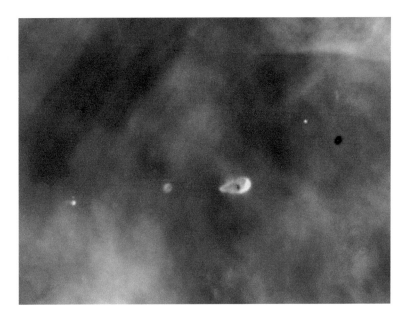

Fig. 1.1 A Hubble Space Telescope view of a small portion of the Orion Nebula reveals five young stars. Four of the stars are surrounded by gas and dust trapped as the stars formed but were left in orbit around the star. These are possibly protoplanetary disks, or proplyds, that might evolve on to agglomerate planets (Image by C. R. O'Dell/ Rice University/NASA/ ESA. Used with permission)

into the region now occupied by the Kuiper Belt." Jewitt compares the dynamic to a giant snowplow moving outwards and picking up snow on the front of the plow. The KBOs all get stuck on the resonance as the resonance moves outward with Neptune. This may be the most radical concept to arise in Solar System studies in the last 20 or 30 years – the notion that the planets were not always where they are now. A decade later, the revelations triggered a theory called the Nice (pronounced "niece" and named after the French city) model, in which the planets' orbits are not only moving – the sizes of the orbits are changing – but sometimes the planets actually pass into a resonance with each other, so that Jupiter and Saturn might get into a mean motion resonance where the ratio of the orbits is 2:1 or 5:2. If that happened, the Nice model argues that the shifting arrangement would really shake up the Solar System and give rise to many of the phenomena that observers see today.

"People like the more extravagant models," Jewitt comments. "People like the Nice model because it's very pictorial and graphic. There's another one called the Grand Tack, where Jupiter and Saturn march in toward the Sun and Saturn pulls Jupiter out at just the right moment."

If the Grand Tack is on the right track, Jupiter robbed Mars and its surroundings of icy, planet-building material – asteroidal stuff that formed outside the snow line – sending it toward the inner system. The most water-rich asteroids are the C types (some of which are up to 20 % water, mostly locked into minerals). Thanks to Jupiter's gravity and its migration in the early Solar System, about one out of every 100 C-type asteroids scattered into the outer fringes of the Asteroid Belt. But for each one of those, at least ten spiraled sunward, delivering water to the terrestrials. This cosmic cocktail would eventually become Earth's oceans.

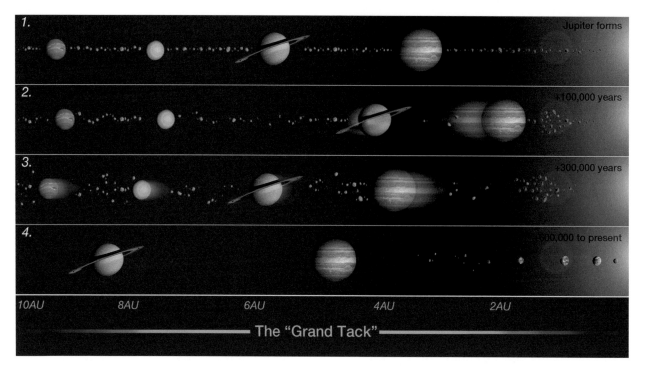

1.
2.
3.
4.

Jupiter forms
+100,000 years
+300,000 years
600,000 to present

10AU   8AU   6AU   4AU   2AU

The "Grand Tack"

Fig. 1.2 The Grand Tack model of Solar System formation. Jupiter and Saturn migrate inward and then out again, scattering and mixing the rocky (brown) and icy (blue) asteroids. Column 1: Jupiter and Saturn form within 6 AU of the Sun; Column 2: Jupiter migrates inward – with Saturn following, disrupting the asteroids. Column 3: Saturn's gravity pulls Jupiter back toward the outer system, while the orbits of Uranus and Neptune shift; icy bodies are cast inward. Column 4: Jupiter and Saturn settle into their modern locations, having cleared out most asteroids and sent the remnant sunward toward the terrestrial planets (Note: Planets and distances are not to scale.) (Diagram by the author, adapted from exoclimes.com/Professor Frédéric Pont)

The Grand Tack version of planetary history has the advantage of explaining the diminutive size of Mars, the structure of the Asteroid Belt and the birth of terrestrial seas at the hands of incoming space rocks. But like high fashion, planetary science comes in trends that fall in and out of favor. "We have these infatuations," Jewitt contends. "These are all really nice models, but we don't really know what happened. We don't know if they represent reality." The Grand Tack and the Nice model are both steps in understanding, incremental advances in our mapping of the Solar System's shifting past. And where once we had eternally concentric circles, we are coming to view the system as a dynamic, changing arrangement in both planetary location and scale (Fig. 1.2).

Once the planets and asteroids had settled down, things became more organized in the outer system as well. Beyond the arena of marching planets and wandering asteroids, the Kuiper Belt came to serve as the reservoir of short-period comets such as Halley's and 67P/Churyumov-Gerasimenko. Even the distant KBOs orbit in elongated ellipses rather than the fairly circular orbits typical of the planets. Pluto travels around the Sun two times for every three circuits made by Neptune in one of those famous resonances. Pluto's is a 3:2 resonance, and nearly a quarter of all known KBOs are in this specific orbital relationship. These objects are referred to as Plutinos. Researchers estimate that some 1400 Plutinos have diameters greater than 100 km. This adds up to just a few percent of the total material in the Kuiper Belt. Though they are small in size, they are mighty in number. But to the first observers, they were unknown.

BEGINNINGS

Planet-hunting was all the rage in the 1700s; there was more to find out there than anyone had suspected. In the previous century, Galileo had shaken the worlds off their securely ordered pedestals by discovering four moons wheeling around Jupiter, circling something besides Earth (lending credence to the Copernican model of a Sun-centered Solar System; everything did *not* revolve around Earth). These attendants were star-like compared to the disk of Jupiter, and no one had a clue as to the titanic size of Jupiter itself. The planet would swallow a thousand Earths with room to spare. Those points of light turned out to be respectable on a planetary scale as well, with mighty Ganymede checking in as the largest moon in the Solar System. Its diameter exceeds that of the planet Mercury, blurring the lines between planet and moon. Not bad for a little star buzzing around Jupiter.

About 5 years later, the Dutch astronomer Christiaan Huygens found another satellite circling Saturn. As with the Galileans, Titan's true size – planetary in scale – would not be appreciated for over a century.

In 1758, Edmund Halley's forecast of the return of a comet came true,[1] and suddenly these mysterious harbingers of doom seemed a bit more predictable and mundane. Observers debated cometary distances and scale. The ancient Greeks believed the ghostly travelers to be within the atmosphere. They reasoned that the celestial phenomena must, therefore, be small, small enough to fit inside that celestial sphere where the birds flew and clouds drifted. Comets turned out to be much farther away and larger than ancient scholars suspected.

The year 1786 brought the apparition of another cosmic visitor, Comet Encke, first seen by Pierre Mechain using coordinates computed by Johann Franz Encke from an earlier apparition. It was the first periodic comet found after Halley's, confirming that such a thing was not a fluke. The second person to observe Comet Encke after Mechain was the talented astronomer Caroline Herschel. Twenty-two years later, Herschel's brother William spotted the first of the "new" planets, Uranus, unknown to the ancients but a world nevertheless. The scale of our neighborhood was expanding (Fig. 1.3).

Within this new onslaught of planets and moons and comets, something seemed to be missing. A great void spanned the distance between Mars and Jupiter, and it was a gap that didn't seem to fit. The planets appeared to orbit the Sun at predictably increasing distances. Johann Daniel Titius and Johann Elert Bode published their "Titius-Bode law" describing this planetary blueprint. Their calculations implied that a planet should be somewhere in between Mars and Jupiter, and another beyond Saturn. When Herschel found Uranus at the prophesied distance, a search was inspired to discover the missing planet outside the orbit of Mars. On the first day of 1801, Italian astronomer Giuseppe Piazzi thought he had found the wayward world in about the right location. He had discovered the asteroid 1 Ceres.[2] While not a planet, Ceres was one of a family

[1] Sadly, Halley did not live to see the confirmation of his prediction. He died 16 years before the comet's return.

[2] All asteroid names are preceded by a number indicating the order in which they were found. 1 Ceres was the first asteroid discovered, while 4 Vesta was fourth to be found.

Fig. 1.3 The apparition of
comet Halley in 1910 served
as spectacular confirmation,
once again, of the comet's
periodic orbit. Its next visit
occurred in 1986 (Image
courtesy of Yerkes
Observatory)

of large space rocks orbiting in a belt between Mars and Jupiter. Thanks in large measure to the "missing" planet predicted by the Titius-Bode law,[3] many more planetoids (planet-like bodies) would be charted soon in the same region.

It seemed to eighteenth-century astronomers that the planetary neighborhood swarmed with tumbling mountains and flying icebergs, and they were right. The asteroids themselves range from the size of rubble piles to the scale of dwarf planets. Ceres – the largest – spans nearly 1000 km across, a third the diameter of the planet Mercury, and comparable in size to Quebec (for a more detailed look at Ceres, see Chap. 2).

ASTEROIDS

Piazzi's discovery of the asteroid Ceres opened up a can of worms. Rather than finding a planet neatly ensconced in an orbit between Mars and Jupiter, he had discovered the first member of an entire clan of mini-planets, rocky and metallic objects orbiting the Sun in a flattened cloud now known as the Asteroid Belt. The asteroids are not confined to this region; some wander in close to Earth and even cross our orbit sunward, while others venture beyond the realm of Jupiter. But the majority, and the largest, restrict themselves to the main belt. Here, more than 200 asteroids measure larger than 100 km in diameter. Researchers estimate that the main belt holds more than 750,000 asteroids spanning diameters greater than 1 km. Still, they cover so much space that the average distance between them is about 1 million km, two and a half times the distance between Earth and the Moon. Travelers to the outer Solar System will never need to dodge swarms of rocks, a la *Star Wars*.

[3] While the Titius-Bode law accurately predicted the location of Uranus, it did not explain the lack of a planet between Mars and Jupiter, nor did it predict the location of Neptune, which does not fit its pattern. Today, the Titius-Bode law is generally considered little more than a numerical oddity.

An astronaut backing away from the Solar System would not see a flat disc of asteroids between Mars and Jupiter, but rather a great swarm of asteroids in a doughnut shape surrounding the Sun, stretching from beyond the Martian orbit to the regions not quite out to Jupiter, with a few stragglers on either side.

Some asteroids are large enough to create a really bad day should one hit Earth. But the asteroids hold fascination far beyond our species' self-preservation. They are the leftovers of the original building blocks of the Solar System, the flotsam and jetsam of the planetary system's formative years. Within them may lie untold secrets of our early planetary history.

Astronomers wondered whether the asteroids were the remains of a failed planet, or a world perhaps destroyed in a cosmic catastrophe. But it turns out that if all the material of the Asteroid Belt were combined, it would only add up to a body smaller than Earth's Moon. What is it doing there? Computer models suggest that the gravity of immense Jupiter disrupts the regions sunward of it, preventing the formation of a planet so close by. Nevertheless, mysteries surround the makeup of the asteroids. Some contain olivine and other materials that typically form under high pressures within the hearts of planets. In a process called differentiation, a planet or moon heats up enough to cause heavy metals to sink to the core, while lighter materials form upper layers and crust. This happens only in fairly large bodies that have copious amounts of heat from their formation or from radioactive elements in their interiors. This is another consequence of size: the larger a planet or moon, the more heat it has to begin with, and the more radioactive elements it has inside, the longer that heat will last. Scientists estimate that some 100 ancient asteroids were large enough to have differentiated, resulting in the rare olivine-rich rocks scattered throughout the belt today.

The mystery is not that there is olivine out there, says asteroid expert Dan Durda of the Southwest Research Institute. Instead, he says, "The big question is where is all the *missing* olivine? The 'standard' story goes like this: We see asteroids that are usually interpreted to be the exposed iron-nickel cores of differentiated asteroids. If they are, then that implies that somewhere there ought to be the mantle and crust material as well. The crust material is easy to account for in the basaltic achondrite asteroids/meteorites that we see. But then, where are all the olivine asteroids that would be from … those differentiated asteroids?" So, the asteroids seem to display chunks of the deep interiors of ancient bodies, but the remnants of their crust – the outer layers – can't be found out there. The Asteroid Belt still poses many such fundamental questions.

Although the asteroids of the Solar System were probably never entire planets unto themselves, the main belt has some impressive inhabitants. As Piazzi might proudly point out, the largest asteroid, and first to be discovered telescopically, is the asteroid Ceres. The Dawn spacecraft went into orbit around Ceres in April of 2015. But the first asteroid to be imaged up close was the stone/nickel-iron worldlet Gaspra, encountered by the Galileo spacecraft on its way to Jupiter. Gaspra's irregular shape, $18 \times 11 \times 9$ km, is reminiscent of a shark. Two years later, Galileo dashed through the Asteroid Belt again, flying by the 54-km-long wanderer Ida. The spacecraft spied tiny Dactyl, a 1.2-km moon, next to Ida, the first asteroid moon seen first-hand. The discovery turned out to be not a unique situation. At last count, 244[4] asteroids have companions of their own. Since Galileo's flight, many spacecraft traveling in the outer Solar System have been targeted to fly by asteroids on their way to their destinations.

Within 6 years of Piazzi's discovery of Ceres, the three next-largest planetoids were found. Second on the family tree of asteroids is the potato-shaped 2 Pallas, some 550 km long. 3 Juno was discovered next. Although smaller, its surface is much brighter than the dusky Pallas or the next-smallest, Vesta. Juno is quite irregular, with dimensions of roughly $320 \times 267 \times 200$ km.[5] Pallas and Juno are only slightly larger than their sibling 4 Vesta. (For more on Vesta and Ceres, see Chap. 2).

The main belt asteroids are broken into an assortment of families according to type and location, including Eos, Gefion, Vestian, Flora, Koronis, Hungaria, Adeona, Baptistina and Eunomia. Meteorites (meteora that hit the ground) and asteroids (meteors that are boulder-sized or larger) from the Vestian group are associated with the asteroid Vesta, and asteroids in the Gefion group probably have their origins in Ceres (Fig. 1.4).

Recently, the media has focused on Earth-crossing asteroids and the dramatic aspects of exploration or of a possible Earth impact. The concept has provided fodder for such Hollywood blockbusters as *Deep Impact* and *Armageddon*.

[4] As of September 20, 2014. This includes 101 confirmed and 83 probable. See William Robert Johnson's summary at http://www.johnstonsarchive.net/astro/asteroidmoons.html#2.

[5] With an uncertainty of ±6 km.

Fig. 1.4 An assortment of comets and asteroids to scale. Each square on the tablecloth is 1 km across (Art by the author.)

At the outer fringes of the main planetary system, beginning at the orbit of Neptune, floats another doughnut of material. This one is 20 times as wide as the Asteroid Belt and 200 times as massive. It is called the Kuiper Belt, after one of the astronomers who theorized its existence. This region extends from 30 AU at Neptune to 50 AU, far beyond our main family of planets. It is the place left behind, the dregs of the solar nebula that remained at the outskirts as the inner solar nebula warmed and condensed into the planets and moons we see today. While the asteroids consist mostly of stone and metal,[6] members of the Kuiper Belt are comet-like, rich in ices of water, methane, ammonia and other volatiles. Many of the comets that frequent the inner Solar System (like comet Halley) come from this region.

Before researchers knew of the Kuiper Belt, our Solar System seemed an orderly affair, says astronomer David Jewitt. "When I started out in the 1980s, there was no discussion of a Kuiper Belt. It didn't exist. There was no talk of it being something to look for. Back then, the idea seemed very strange that the outer Solar System was as empty as it looks, given that the inner Solar System was so full of asteroids and comets and planets and stuff. But beyond Saturn, we just knew of Uranus and Neptune and Pluto." That contrast was what triggered the search for other bodies in the outer Solar System. "We would have been delighted to find anything beyond 10 AU," says Jewitt, "because nothing had been found out there." Astronomers soon realized that it was surprisingly easy to find asteroids in the main belt, and also quite easy to find Trojan asteroids of Jupiter at a distance from the Sun of 5 AU, but the telescopes of the time simply could not detect anything at greater distances.

Meanwhile, a Canadian/MIT group, Martin Duncan, Thomas Quinn and Scott Tremain, wrote a paper[7] speculating on the origin of the short-period comets. The authors demonstrated that an earlier comet idea could not be correct. The earlier theory suggested that all comets come from a vast distant sphere of comets at the very fringes of the Solar System (see the comet section below). The idea was put forth by Dutch astronomer Jan Oort, and it turns out that he was partially correct. If there were a great cloud of comets in random orbits outside of the Kuiper Belt, those comets would come from all different directions. In fact, there is a direct correspondence between the haphazard orbits in the Oort Cloud and the random orbits of the long-period comets. All the large planets and asteroids tend to go around the Sun in a counter-clockwise direction when viewed from above the north pole. This is known as prograde motion. Bodies that follow retrograde paths (in the opposite direction) are rarer. But the long-period comets echo the random motions of Oort Cloud objects: equal numbers of prograde and retrograde, northern hemisphere and southern hemisphere, and so on. The short-period comets are another matter, says Dave Jewitt. "They know where the midplane of the Solar System is; they're in a sort of disk that's only about 30° thick. Oort suggested that the gravity of Jupiter converted the long-period comets into the short-period ones that we see with small inclinations. He said that in 1950, when nobody had computers. But eventually Duncan, Quinn and Tremain did this calculation in

---

[6] Some, like Ceres, also have a large component of ice.

[7] *Astrophysical Journal*, 328: L69–L73, 1988 May 15.

which they basically established that you could not capture short-period comets from the long-period population."

At the time, Jewitt guessed that the Solar System's emptiness beyond Jupiter was an illusion caused by the limitations of the instruments. "You double the distance and the object is sixteen times fainter," Jewitt points out.

Jewitt's search was years in the making. The first object discovered was called 1992 QB1. Jewitt found many dozens of objects over the next few years. Astronomers used the paths of those objects to map out the structure of the Kuiper Belt. As the maps took shape, mysteries unfolded. Researchers thought that this region of the Solar System would be much like the disk from which the entire Solar System – the rest of the planets – accreted. We know that disk was very thin. But it turns out that the Kuiper Belt is very thick – even thicker than the Asteroid Belt – more like a doughnut than a sheet of paper. Scientists now know the reason: the Kuiper Belt has been stirred up – or, as dynamicists put it, excited – by an unknown force or forces. This "puffing up" of the Kuiper Belt may have been caused by distant large planets or by passing stars. Jewitt and his colleagues have also discovered that the Kuiper Belt has a complex structure of orbiting objects.

Some Kuiper Belt objects may rival the planet Mercury in size. Pluto was the first to be seen, when Clyde Thombaugh spotted a distant, starlike object moving across his photographic plates in 1939. His discovery was heralded as the ninth planet of the Solar System. Pluto held that title until recently, but it was demoted because, among other reasons, it had a whole lot of company. Many other objects have been found with better telescopes. In addition, neither Pluto nor any of its siblings of the outer Solar System orbited in the same stately plane as the major planets. Now, researchers see Pluto as one member of the throng of billions of rocky and icy planetoids known as Kuiper Belt objects, or KBOs. Sedna, discovered in 2004, lies so far out that it takes 105 centuries to circle the sun. Its oblong orbit, ranging from 13 to 135 billion km from the Sun, is fairly typical of the odd orbits of Kuiper Belt members.[8] Another, Eris, is nearly a twin in size to Pluto, and a third, 2000-km-long Haumea, is distinctly football-shaped with two known moons. There may be thousands of planet-sized objects lurking out in the Kuiper Belt (see Chap. 2). Current estimates put the number of objects larger than 100 km diameter at 100,000, with up to ten billion larger than two km in diameter. For each asteroid in the main belt, there may be a thousand in the Kuiper Belt. And, as Jan Oort might tell you, our solar neighborhood doesn't end there.

COMETS

Although grander in scale across the sky, the heart of a comet is a much smaller affair than an asteroid. While its foggy coma and gas/dust tail may stretch for upwards of 150 million km, a comet's nucleus, source of that glowing spectacle, is only a few km across, smaller than many of the asteroids seen up close so far. While asteroids are collections of rock and metal, comets are rich in volatiles such as water, hydrogen cyanide, carbon

[8] Sedna is actually so far out that it may be an inner member of the Oort Cloud.

monoxide, ammonia and methane. However, some asteroids also possess volatiles, and may be burned out comets. The distinction between the two is sometimes as hazy as a comet's tail.

When the path of an icy KBO is nudged by the gravity of one of the giant planets, it may coast inward, toward the Sun. As it warms, it develops a cloud that streams off in a long tail under the pressure of the solar wind. It becomes a comet. Comets from the Kuiper Belt are called short-period comets, as they travel in orbits lasting a few years. But occasionally, comets come in from a very long distance away. These long-period comets are seen only once in a lifetime, returning after hundreds or thousands of years. Comets with this type of long orbit come from an icy reservoir farther out than the Kuiper Belt, the Oort Cloud. These cosmic icebergs inhabit a shell of dust and ice enfolding the Solar System at distances up to one light-year away. Rather than the doughnut shape of the asteroids and Kuiper Belt, the Oort Cloud is a vast bubble encircling the entire solar neighborhood. Both short- and long-period comets play an important part in our understanding of the nature of our Solar System and others.

To that end, the European Space Agency embarked upon the Rosetta comet mission. Rosetta carefully approached the short-period comet 67P/Churyumov-Gerasimenko (also called P67) in 2014, settling into an uneasy orbit around the tiny, nearly gravity-free ice ball.[9] Its goal: to accompany P67 into the inner Solar System, observing the nucleus as it vaporized and became an active comet. To remain beside the comet nucleus, Rosetta would need to continually correct its orbit (Fig. 1.5).

On November 12, 2014, Rosetta deployed an ambitious lander called Philae. Philae is named after the Egyptian obelisk that, along with the

[9]In the case of bodies as small as a comet nucleus, orbit is a loose term. While Rosetta is held precariously within the weak gravity of P67, it travels in a series of triangles, firing its engines three times for every circuit around the comet to remain close.

Fig. 1.5 Left: Cheops boulder (arrowed) on Comet Churyumov-Gerasimenko compared to the Great Pyramid of Cheops and, more appropriately, to St. Basil's Cathedral in Moscow (right) (Comet images © ESA/Rosetta/MPS for OSIRIS Team MPS/UPD/LAM/IAA/SSO/INTA/UPM/DASP/IDA; aerial Giza plateau WWII image courtesy of Ellie Crystal; St. Basil's photo by the author, digitally remastered)

Rosetta Stone, helped Jean-Francois Champollion break the code of the Egyptian hieroglyphs. Scientists hope that Philae and its mother ship will help to decode the history of planetary formation.

Philae landed softly on the surface of the comet, but its ground-hugging harpoons failed to deploy. Under the comet's weak gravity, the plucky lander bounced off the surface and had a "hang time" of approximately 2 h, reaching an altitude of at least 1 km. It bounced a second time, but settled to the surface about 6 min later. The lander rests at a 30° tilt, wedged against an outcropping in rugged terrain. Shadows from the surrounding landscape obscured Philae's solar panels, so that after 60 h of surface operations, the lander succumbed to low energy levels, putting itself into hibernation. But its brief, furious investigations from the surface included drilling and measurements of the surface consistency, which turned out to be similar to solid water-ice (rather than the expected fluffy texture). The surface was so hard that the drill was unable to deliver samples to its internal ovens, but the lander did sample the comet's atmosphere, discovering organic molecules including hydrogen and carbon. The lander returned data on other gases, magnetic fields, and temperature (−153 °C/−243 °F) and returned a spectacular panorama and microscope views, the first from the surface of a comet. Using a radar system in concert with the Rosetta orbiter, Philae was able to probe the comet's core, which turns out to be more porous than the surface.

On June 13, 2015, after an ominous 7 month silence, flight controllers at ESA's center in Darmstadt received an 85-s transmission from the prodigal lander. The data had to be relayed by the orbiting Rosetta. The brief transmission signaled that the lander was in good health and had recharged its batteries enough to come out of safe mode. Philae had actually come out of hibernation some time before June 13, but had been unable to contact Rosetta before then. The lander was operating with 24 watts of electrical power at −35 °C (−31 °F). Controllers had only sporadic communication until July 9, when the lander transmitted 12 precious minutes of measurements from its CONCERT (COmet Nucleus Sounding Experiment by Radiowave Transmission) experiment. As of this writing, two-way contact with the Philae lander has still not been possible, but heroic attempts continue. Meanwhile, the intrepid Rosetta orbiter continues its survey of Comet 67P/Churyumov-Gerasimenko, following along as the cosmic iceberg loops its way around the Sun on its way back to the outer reaches of the Solar System.

What has Rosetta learned? Before the Rosetta mission, astronomers believed that comets contributed most of the water in Earth's oceans. But Rosetta demonstrated that this is not the case. Rosetta researchers measured the levels of hydrogen isotopes firsthand. Comet 67P/Churyumov-Gerasimenko – and presumably most other comets – has a greater deuterium[10] to hydrogen ratio than that found in Earth's oceans. This suggests that if comets supplied water to Earth's oceans, their contribution was minimal. It now appears that most of Earth's water came from ices within the Asteroid Belt. In the early Solar System, asteroids fell in toward the inner Solar System, depositing minerals and ice upon the terrestrial planets. Most comets, however, remained beyond what is now the orbit of Jupiter.

---

[10]Deuterium is an isotope of hydrogen that weighs twice as much. It is used to calculate how much ancient water a planetary body had during its development.

The region nearest the Sun, where the terrestrial planets formed, was too hot for water or gas to condense out as ice. Planets there gathered into the silicate, rocky bodies we see today – Mercury, Venus, Earth, Earth's Moon, and Mars. But farther out in the great disk of the solar nebula, temperatures were cold enough to form ice. The line between the two regions is called the "snow line." Beyond that line, smaller bodies congregated into spheres of ice and rock. The gas giants, with their immense gravity, pulled hydrogen and helium directly from the nebula around them, ending up as massive globes of hydrogen, helium and ammonia with metallic cores. Far from the Sun's warmth and blustery solar wind, these planets retained materials virtually unchanged from the very birth of the Solar System. Like them, the comets represent a sampling of the solid materials in the outer, primordial disk. Their ice and dust retain a precious record of the early Solar System's nature. The Rosetta mission has now given us an up-close glimpse at what that nature was.

Comet P67/Churyumov-Gerasimenko is no rock or iceberg. It is as fluffy as cork, with a porous interior probably filled with small hollows and caverns. The coma surrounding the nucleus, a cloud of dust and gases from the comet's interior, morphs and changes daily. With three times as much dust as gas, the cloud is a swarm of microscopic particles mixed with a few large chunks up to a meter across. Jets – made mostly of water vapor – billow out into the coma. Most of the plumes stream from the neck area, the narrow middle section of the comet. The surface, expected to be icy, is in fact very dry, with a crust of desiccated organic materials. This surface is a complex mix of carbon-hydrogen or oxygen-hydrogen chemicals. Small outcrops or pebbles of water-ice peek out here and there, but Rosetta did not see any patches of ice (Fig. 1.6).[11]

[11] For detailed articles, see the January 23, 2015, issue of *Science*.

Fig. 1.6 *The nucleus of Comet P67 compared to Mt. Fuji (Art by the author, photo via Wikipedia Commons, https:// commons.wikimedia.org/ wiki/File:FujiSunriseKawaguc hiko2025WP.jpg)*

P67/Churyumov-Gerasimenko and other short-period comets like it are rich in carbon and other building blocks of life. This makes the comets a prime target in our search for the origins of life and the processes that led to the planetary system we live in today.

COMET AND ASTEROID ENCOUNTERS

Rosetta was not the first spacecraft to encounter a comet or asteroid. Below is a timeline of such expeditions.

Spacecraft	Agency	Encounter date/s	Target/s	Comments
ICE	ESA/NASA	1978	21P/Giacobini-Zinner	Flyby
		1986	Comet halley	Distant observations
Giotto	ESA	1986	Comet halley	Flyby
		1992	26P/Grigg-Skjellerup	Flyby
Vega 1 and 2	IKI (USSR)	1986	Comet halley	Flyby
Suisei	JAXA (Japan)	1986	Comet halley	Flyby
Sakige	JAXA (Japan)	1986	Comet halley	Flyby
Galileo	NASA/JPL	1991	951 Gaspra	First asteroid flyby
		1993	243 Ida/Dactyl	First closeup of asteroid moon
NEAR shoemaker	NASA/APL	1997–2001	253 Mathilde	Flyby
			443 Eros	First asteroid orbit/landing
Deep space 1	NASA/JPL	1999	9969 Braille	Flyby
		2001	19P/Borelly	Flyby
Cassini	NASA/ESA/ASI	2000	2685 Masursky	Distant flyby
Stardust	NASA/JPL	2002	5535 AnneFrank	Flyby
		2003	81P/Wild2	Comet sample return
		2011	Tempel 1	Deep Impact follow up
New horizons	NASA/APL	2006	132524	Distant flyby
Deep impact	NASA/JPL	2005	9P/Tempel	Comet impactor/flyby
Deep impact/ EPOXI		2005	103P/Hartley	Flyby
Hayabusa	JAXA	2005–2010	25143 Itokawa	First asteroid sample return
Dawn	NASA/JPL/ESA	2011–2012	4 Vesta	In orbit for 14 months
		2015	Ceres	Second orbital mission/ first craft to orbit two bodies
Chang'e 2	CNSA (China)	2012	4179 Toutatis	Flyby
Rosetta/Philae	ESA	2008	2867 Steins	Flyby
		2010	21 Lutetia	Flyby
		2014	67P	First comet orbit/first comet landing
Projected missions				
Hayabusa 2	JAXA	2018	993JU3	Sample return 2020
OSIRIS-Rex	NASA	2019	Bennu	Launch 2016
				Sample return 2023
New horizons	NASA/JHUAPL	2019	PT-1	Kuiper Belt object
Asteroid redirect	NASA	late 2020s	Possibly 2011MD	Manned mission to bring asteroid into lunar orbit

ASTEROIDS, COMETS AND YOU

A flock of distant rocks and icebergs circling the Sun may seem to have little to do with everyday life, but what if one of these things came down in our own backyard? Such an event took place near the Russian city of Chelyabinsk, a city of 1.3 million on the border of Siberia. On the morning of February 15, 2013, shortly after dawn, a bright, sunlike fireball tore across the sky from the southeast at 9:20 local time. The object creating all the drama exploded before it got to the ground, fragmenting at a location some 40 km south of the city at an altitude of just over 23 km. Fragments of the bollide, or exploding meteor, continued down to shower the area surrounding Lake Chebarkul, where some have been recovered. Many drivers in Chelyabinsk equip their cars with dashcams (dashboard-mounted cameras), and some captured the meteor's dramatic flight. Thousands of people witnessed its spectacular flash. Bystanders were drawn to windows to see the source of light and the glowing trail stretched through the sky. The several-minute-delay in the arrival of the shockwave meant that many were beside windows when the blast shattered glass. In the end, 1491 people reported serious injuries from temporary blindness to mild ultraviolet burns to lacerations from flying glass. About 7200 structures in six cities, including homes, factories and warehouses, were damaged by the meteor's shockwave. Inhabitants of Chelyabinsk reported smells similar to sulfur or gunpowder beginning an hour after the bollide's flight, and lasting into the evening.

The Chelyabinsk incident raised a red flag, according to former Apollo astronaut Russell Schweickart. Schweickart is cofounder of the B612 organization, a non-profit group dedicated to the detection and mapping – and protection of Earth from – near-Earth asteroids. "The most important thing that we learned from Chelyabinsk was that an object of that size – ~18 m in diameter – could do considerable damage on the ground. It didn't kill anybody, but it certainly wasn't far from doing that. The town of Chelyabinsk was some 40 km to the side of the ground track, and it came in at a very shallow angle so that the overall energy was deposited over a fairly large area. What that said is objects of a smaller size than we thought can do pretty serious damage."

When it comes to falling cosmic rocks and their effect, size really does matter, Schweickart contends. "The energy of an object coming in to impact Earth has to be dissipated somewhere in Earth's environment. Partly it goes into the atmosphere. If it reaches all the way to the ground, it ends up in the impact itself. The energy that it brings in is one-half the mass of the object times the square of its velocity. If you have a 40-m diameter object, an 80 mm object will be twice the diameter, but the mass is eight times larger. The energy released would go up proportionally."

Chelyabinsk was not the first of its kind. On June 30, 1908, another such airburst occurred in Siberia. Like the Chelyabinsk incident, the impactor didn't make it to the ground. Instead, a meteor or comet nucleus

[12] Estimates range from 10 to 15 megatons.

Fig. 1.7 Left: The Tunguska event, depicted here by artist Don Davis, resulted when a meteor exploded in the air above Siberia, Russia (© Don Davis, used by permission). Right, above: The incident at Tunguska leveled trees over 40 km away from the blast's center. (Leonid Kulik expedition, public domain). Right, below: The epicenter of the Tunguska airburst today is a marsh in the Siberian wilderness (Image credit: Tungus1908, via Wikipedia commons; https://commons.wikimedia.org/wiki/File:Tunguska_epicenter.jpg)

detonated 6–12 km above the Tunguska River in a nearly uninhabited area. The airburst caused widespread havoc, leveling some 80 million trees and killing wildlife across an area of 2150 sq. km. Exploding with an estimated power 1000 times greater than the atomic bomb dropped over Hiroshima,[12] the remote blast was witnessed by many. Semyon Semyonov was working at the Vanavara Trading Post on the bank of the Tunguska River. He reported in a letter to Moscow's Kulik scientific expedition – specially formed to investigate the incident – that, "I was sitting on my porch facing north and then in an instant a fiery flare took shape in the northwest from which there came so much heat that it was impossible to remain sitting – my shirt nearly burst into flame while still on me. And what a glowing marvel it was! …at that moment the sky slammed shut, and a mighty crash resounded and I was thrown about three sazhen's to the ground. For a moment I lost consciousness, but my wife, running out, brought me back into the hut." Others reported thunder like cannon fire, widespread deaths of reindeer herds, and a "fiery sphere" in the sky (Fig. 1.7).

The Chelyabinsk bollide was so small, and its approaching path so close to the Sun's location, that astronomers were unable to see it coming. But larger asteroids cross Earth's orbit fairly frequently. Just 16 h after the meteorfall at Chelyabinsk, observers were ready for an unrelated intruder that they had already been following. Known as the asteroid 367943 Duende, or $2012DA_{14}$, the asteroid quietly coasted through the Earth/Moon system, upstaged by the unscheduled Russian meteor before it.

Fig. 1.8 Asteroid Duende (at left) compared to the Chelyabinsk meteor, floating peacefully within the Heinz Field sports stadium (Art by the author; Heinz Field photo courtesy of Wikimedia/ Bernard Gagnon, https:// commons.wikimedia.org/ wiki/File:Heinz_Field02.jpg)

Measuring some 30 m across and weighing 40,000 m. t., the Duende space rock could surely have made it to the ground had it been on a slightly different path, causing far more extensive damage than the Chelyabinsk airburst. On this pass, the asteroid came within 27,000 km of Earth's surface, a distance equivalent to just over two Earth diameters, closer than the orbits of our communications and weather satellites. (The Moon orbits at a distance of 30 Earth diameters away) (Fig. 1.8).

What would happen if an asteroid akin to 367943 Duende dropped out of our skies? Such an event did take place in Arizona a few tens of thousands of years ago. We can gain an appreciation of the power pent up in these nearby asteroids and comets by revisiting the formation of the Barringer Meteor Crater.

It may have been a perfectly nice day, with deer, mammoths and mastodons grazing happily within the grassy juniper-pinyon woodlands of what is now central Arizona. The great ebb and flow of glaciers of the Wisconsin interstadial period continued to the north, but here it was warm. In the distance, on the open plain, herds of bison likely ruminated on a variety of grasses at the edge of conifer forests. Ground sloths, wild horses and camels may well have browsed at the base of Ponderosa pines. But the idyllic Pleistocene landscape[13] was about to see an abrupt, terrifying change.

[13] The Meteor Crater impact has been dated with several independent techniques, including thermoluminescence, cosmogenic ^{36}Cl techniques and exposure ages based on *in situ* production of ^{10}Be and ^{26}Al. These independent studies all indicate an event at 50,000 years ago, ± 2350 years (average).

No matter the time of day or night,[14] the southern skies would have bloomed with an untimely sunrise, as a 110,000 ton[15] pile of metal screamed through the firmament. The cosmic intruder was traveling between 12.8 and 20 km/s (or 28,600–45,000 mph). At that breakneck speed, the meteor would have crossed the distance from Los Angeles to San Francisco in ~31 s.

Its landfall triggered an explosion of roughly 2.5 megatons, the power of 2½ million tons of TNT. Within 3 km of the crater site, 2000 km/h winds tore across the landscape, frying everything in their path. Trees bowed radially from the explosion, flattened in a great circle as far away as 19 km. Hurricane-force winds blasted the woodlands up to 40 km distant. Any animals within 10 km of the impact site were fatally burned, and killed or injured by flying debris out to 24 km.

A planet like Earth has a way of healing itself, of covering over blemishes and weathering away wounds. Life returned to the impact site in a matter of decades. The Diablo Canyon Arroyo, in existence before the meteor's arrival, continued its flow, and the great impact depression itself filled with water, becoming a lake. The water-filled crater provided habitat for new types of creatures in the area, and the forests grew anew, morphing into the landscape we see today.

The Barringer Meteor Crater may be the best-preserved impact feature in the world. It spans 1.2 km across. Its bowl dives some 180 m deep, encircled by a hummocky rim rising 30–60 m above the rolling plains around it. The nickel-iron meteorite that came screaming out of the sky to create the crater may have been as large as 60 m in diameter, as far across as three boxcars end to end. Nearly half of the meteor vaporized as the projectile seared through the atmosphere. At impact, the meteor penetrated to a depth equivalent to its diameter or a little deeper, transforming into molten rock and metallic vapor. Less than a tenth of it survived as solid fragments (Fig. 1.9).

While the flambéed elk and braised bison may not have appreciated the impact's magnitude, its scope was remarkable. 175 million tons of rock sprayed into the sky in a mushroom cloud that could have been seen hundreds of miles away. Along the outside edge of the crater, the blast draped a mirror image of the rock layers upon itself, opening the ground like a giant rose blossom. This unfolding of layers is seen at impact craters throughout the Solar System.

B612's Schweickart explains. "The one that hit in Winslow, Arizona [forming the Barringer Meteor Crater] was an iron-nickel object. Something like that will come in, and it does not break up as easily in the atmosphere as a stony object would. Something like the object that made the Meteor Crater in Arizona was probably no larger than the object that hit in Tunguska back in 1908, but that one broke up in the atmosphere because of its relatively low density and fractures and things in it… a similar-sized iron-nickel object probably makes it all the way to the surface. There are a few asteroids made of iron and nickel. Those obviously originated in asteroids that were fairly large and had a chance to differentiate,

[14] Due to the meteor's trajectory, researchers estimate that the event occurred in the early morning or late evening.

[15] Estimates have ranged from 5000 to over 500,000 m. t. The figure given is from David Kring's recent calculations, printed in LPI's *Guidebook to the Geology of Barringer Meteorite Crater.*

Fig. 1.9 The meteor that blasted out Arizona's Barringer Meteor Crater may have been as far across as three boxcars (50 m) (Art by the author, model train cars by Matt Sugerman, used by permission)

Fig. 1.10 The Barringer Meteor Crater, seen from the air, is a vast bowl 1.18 km across and 170 m deep (as deep as a 50-story building). Note the Visitor Center, offices and parking lot at lower left (Wikipedia Creative Commons, by Shane Torgerson; https://commons. wikimedia.org/wiki/ File:Meteorcrater.jpg)

so that all of the heavy elements went to the center and then got broken up. You end up with some of the asteroids being iron and nickel, but less than 6 %" (Fig. 1.10).

Are the Barringer meteor impact and Tunguska event incidents that are likely to replay any time soon? Chelyabinsk and Duende give experts pause, and judging by the cratering record of the surfaces of other terrestrial worlds (specifically the Moon, Mercury and Mars), Meteor Crater is a relatively small impact event. As such, it may be a fairly common occurrence on cosmic time scales. Current estimates suggest that a similar impact transpires on the average of once every 1600 years somewhere on Earth, and about once each 6000 years on dry land. As Meteor Crater expert David Kring writes, "This is a time scale that is both meaningful and memorable in terms of human history and should probably be kept in mind when evaluating the hazards of objects in or entering near-Earth space."[16]

One of the most famous impact events took place 65 million years ago and is widely blamed for the final extinction of the Mesozoic dinosaurs.[17] At the end of the Cretaceous Era, dinosaur species were dwindling. The fossil record indicates that populations were declining in number already when a titanic asteroid struck Earth off the coast of what is now the Yucatan Peninsula. The evidence for the "terminal Cretaceous-Tertiary event" (KT event) is both indirect and direct. We can see remnants of the giant crater, submerged in the Gulf of Mexico and draped across the Mayan heartland, by orbital gravity maps. These maps are assembled by carefully tracking the orbits of satellites. Their orbits vary due to subtle differences in Earth's density, providing a picture of what's under the surface. The impact feature is some 110 miles across, and is now known as Chicxulub. A line of cenotes, natural sinkholes, rings what used to be the crater rim. These compressed rings of stone resulted from the shockwaves radiating through the peninsula. Strewn across the entire Earth is a layer of iridium, a metal that is rare on our world but common to some meteorites. Below this layer, dinosaurs ruled. Above it, they vanished.

Compared to the Chelyabinsk and Barringer meteors, Chicxulub was a terrifying apocalypse. Equal to 50 trillion Hiroshima atomic bombs, the devastating explosion would have pulverized the asteroid into fine grains that rushed back up into the atmosphere and into space. What remained in the atmosphere would became a globe-encircling shroud of soot. Shocked quartz crystals, indicative of great pressures, would have rained from the sky (such fragments having been found in associated geologic forms related to Chicxulub). Adding to the tumult, the heat of the impact and its fallout would have triggered worldwide flash fires. The smoke would have contributed to the atmospheric haze, blocking the Sun across the entire Earth. Temperatures would have dropped, photosynthesis would have virtually ground to a halt, and the food chain would have collapsed, beginning with the preeminent carnivores such as Tyrannosaurus, and trickling rapidly down in a disastrous ecological cascade of doom.

Chicxulub and Barringer meteor craters serve as cautionary tales for the human race. Astronomers are mounting searches for asteroids that regularly cross Earth's path, and proposals are on the table to send space borne observatories to further chart potential hazards. NASA and other space agencies

[16]"Air blast produced by the Meteor Crater impact event and a reconstruction of the affected environment," by David A. Kring, *Meteorics and Planetary Science 32*, 1997.

[17]Some research points to modern birds as being surviving dinosaurs.

are studying ways of deflecting asteroids. Even the United Nations has a task force dedicated to the threat, and to global education on the subject.

Schweickart's B612 Foundation is on the forefront of the effort. The first step in that effort is to simply detect and catalog the asteroids that cross Earth's orbit. To that end, B612 proposes a unique observatory called Sentinel. "The location of Sentinel would not be in orbit around Earth," Schweickart says. "We're in a solar, Venus-like orbit. We go around faster than Earth, and we're looking outward basically plotting everything that crosses Earth's orbit, the things that will eventually hit Earth." Sentinel's advantage is its faster orbit, lasting less than one Earth year. An observer who is Earth-locked sees objects in Earth-like orbits for only a few years, after which time they disappear for an extended period. Sentinel's interior orbit enables it to track near-Earth objects (NEOs) more frequently, and for longer periods, Schweickart explains.

The job ahead is daunting. Schweickart estimates that there are one million Tunguska-like objects, roughly 40 m across. Of these, about one percent have been detected. "That means we have no idea where 99 % of the Tunguska-type objects are." In the case of 20-m objects like Chelyabinsk, the situation is even more grave. "We've got about ten million of those, and maybe one tenth of one percent of those have been found. In all likelihood, we'll never [catalog] the entire population of Chelyabinsk-like objects. There are too many of them. They're too hard to find." Even finding 40-m objects is a huge challenge that will take combined efforts of space telescopes and ground telescopes, and it will take decades. B612 hopes for a 2018 launch window to begin the next phase of this search, says Schweickart (Fig. 1.11).

To underscore the likelihood of impacts, meteorite experts have logged 188 confirmed impact craters across the face of Earth, ranging in size from a quarter of a kilometer to the 180-km-wide Chicxulub crater draped across the Yucatan. Using projections of how frequently space rocks hit Earth, along with how efficiently erosion and sediment wipes them out, they estimate that some 350 more craters of at least 250-m size remain to be discovered.

Fig. 1.11 Left: The Chicxulub Impactor was at least 10 km (6 mi.) in diameter, and would have dwarfed San Francisco Bay (Photo courtesy Hispalois, Wikimedia Commons: commons.wikimedia.org/ wiki/File:SF_Bay_aerial.jpg). Center: Gravity map of the crater's structure (NASA). Right: The layer of iridium can be found worldwide and coincides with the end of dinosaurs in the fossil record (Photo by the author)

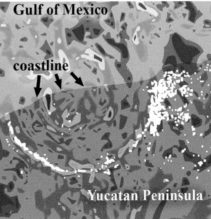

Fig. 1.12 Artist George Solonevich's illustration of three theories of crater formation from The Moon, Our Neighboring World (Golden Press 1959)

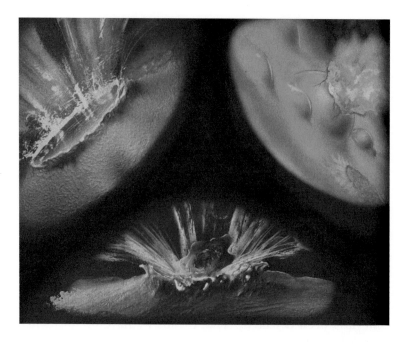

Long before the realization that some of Earth's bowl-shaped depressions originated from impact events, observers struggled to learn the source of the nearby craters scattered across the surface of the Moon. Three leading theories came to the forefront. (1) The volcanic theory proposed that craters were remnants of ancient volcanic eruptions. Volcanic craters were well-known on Earth, and many took on the general appearance of some lunar craters. (2) The bubble theory suggested the Moon began as a molten orb, and gas bubbling from its interior rose to the surface, building domes. These domes either burst or cooled and collapsed, leaving behind only a rim. (3) A third theory, initially not very popular among scholars, was that the surface of the Moon had been pelted by a "flock of giant meteors," leaving behind the impact scars. In the Golden Book *The Moon, Our Neighboring World*,[18] author Otto Binder declares, "Even if the moon was not molten, the great impact would melt stone and splash it aside…[but] Why are no meteoric craters being formed today?" Binder points out flaws in each theory and goes on to conclude, "Like the canals of Mars, this moon mystery may not be settled until scientists land in spaceships someday" (Fig. 1.12).

Of course, theory number three was correct. The Moon shares the same cratered terrain found across the entire Solar System, a rugged landscape sculpted by meteoroids, asteroids and comets. Just what is the nature of nearby asteroids and comets? Let's visit an assortment.

VESTA

[18] *The Moon, our neighboring world*, by Otto Binder. ©1959 by Golden Press, Inc.

If the massive asteroid Ceres (see Chap. 2) is considered a dwarf planet, then the largest true asteroid is Vesta. NASA's Dawn spacecraft settled into orbit around the 525-km planetoid 14 months before heading off to

Fig. 1.13 A section of Vesta's spectacular equatorial canyon system. The largest crater, at lower right, is about 10 km across (Image courtesy of NASA/JPL-Caltech/UCLA/MPS/DLR/IDA)

reconnoiter with Ceres. It found the asteroid to be a rocky, complex world with a differentiated core (where heavy material has settled to the center), a structure much like the terrestrial planets. Its tortured surface plays host to the titanic Rheasilvia Crater near the south pole, whose central peak rises 23 km above the crater floor, a full 2½ times the height of Mt. Everest. Around the dwarf planet's equator, a network of nearly parallel canyons inscribes the cratered landscape. The longest, Saturnalia Fossa, stretches some 370 km long, making it among the longest chasms in the moons of the Solar System. Only Tethys' Ithaca Chasma can compete, and it resides on a moon of Saturn that is twice Vesta's diameter. The asteroid shows signs of geological activity. Pits and flows of material may be telltale signatures of past volcanism. The unique "pitted terrain" resembles vent areas on Mercury and Mars associated with outgassing from the heat of impacts in volatile-rich surfaces (Fig. 1.13).

With evidence from space probes all but ruling out comets as the primary source of Earth's planetary waters, researchers have turned to the asteroids. It's an old idea with new evidence to support it. One such line of evidence comes from meteorites whose source is Vesta. At Woods Hole, geologist Adam Sarafian and his colleagues studied the miniscule amounts of water trapped within apatite, a family of phosphate minerals found in Vestian meteorites. Sarafian's team found that the deuterium (heavy hydrogen) to hydrogen ratio closely matched that of Earth. Their conclusion: the same types of meteors and asteroids that delivered water to Vesta could also have dropped it off on Earth, where it ended up as the oceans on our water-bathed world.[19]

[19] See *Science News* online, November 1, 2014.

The water of Earth resembles the water found among the asteroids far more closely than it does the water in comets studied so far. At least two comets do have deuterium levels close to terrestrial oceans and ancient rock samples. This might tempt some researchers to look to comets as the source of Earthly oceans. But those who study the movements and gravitational perturbations of comets by the planets are not so tempted. It turns out that the gravity of mighty Jupiter tends to shield the inner Solar System from incoming comets. The king of worlds makes it difficult for comets and asteroids to venture into the inner Solar System, as its location in our planetary lineup and its mammoth gravity combine to toss out the interlopers. Dynamically, it is difficult to get enough comets near enough to Earth to make an ocean; Jupiter's gravity simply tosses them out. The few that do make it through don't hang around the terrestrial planets for very long. They either hit a terrestrial planet within a few orbits or get scattered back out into other orbits. Ironically, the same Jovian gravity that keeps comets out may well keep asteroids in. Although Jupiter's disruptive gravity kept a big planet from forming outside of Mars, the giant world tossed much of the planet-forming material inward, toward the Sun and Earth. The desiccated Earth needed moisture to fuel its oceans, and those asteroids may well have served it up, thanks to Jupiter.

Within the family of asteroids, Vesta's size is second only to Ceres. Ceres is a perfect example of an ice dwarf. It is large enough to have become spherical during formation, but it retains even more volatile material than Vesta, most of which is in the form of water-ice. This abundance of water is the norm that we see in many of the smaller bodies of our planetary clan, from the KBOs such as Pluto to the moons of the giant worlds (Fig. 1.14).

Fig. 1.14 Left to right: The two largest asteroids, Vesta (525 km) and Ceres (950-km diameter), compared to the Uranian moon Miranda (470 km) and Saturn's Tethys (1062 km), with Earth's Moon (3475 km) along the base (Vesta and Ceres images courtesy of NASA/JPL-Caltech/UCLA/MPS/DLR/IDA; all other images courtesy of NASA/JPL)

2867 STEINS AND 21 LUTETIA

During its decade-long cruise to orbit comet 67P/Churyumov-Gerasimenko, ESA's Rosetta spacecraft busied itself by flying past several other targets, including the asteroid 2867 Steins. Steins is a rare E-type asteroid, high in albedo (brightness) with a surface made up mostly of material similar to basalts. E-type asteroids are more abundant in the inner Asteroid Belt, becoming rarer with distance from the Sun. The 6.7- by 4.5-km asteroid is irregular, with a flat top and a tapering base that leads to comparisons of a "diamond in the sky." In fact, its shape inspired the International Astronomical Union to establish a guideline that calls for features on Steins to be named after precious stones. Hence, we find craters named Sapphire, Onyx, Obsidian, Aquamarine, and, of course, Diamond. The top surface holds a crater fully 2.1-km across, and many researchers were surprised that the asteroid survived an impact so large compared to its overall size.

Rosetta's second drive-by passed the spectacular asteroid 21 Lutetia. Much larger than any asteroid previously visited, Lutetia turned out to be a mammoth boulder of stone and metal. The spacecraft provided an estimate of Lutetia's density, which turned out to be surprisingly high. Rosetta's flight path carried it much closer to the asteroid than its flyby with Steins, enabling it to see the northern half of the asteroid up close. Some areas of Lutetia were heavily cratered, while others seemed to have been resurfaced in the geologically recent past, leaving fewer craters exposed. The two most heavily cratered regions, Achaia and Noricum, exemplify the most ancient areas on the surface of Lutetia, dating back some 3.4 and 3.7 billion years or more. This makes them nearly as old as the asteroid itself. Some of the craters that densely populate these two regions date back to an early period in the Solar System's history, right after the epoch known as the Late Heavy Bombardment, when there was a rain of meteors and asteroids during the formative years of the Solar System some 3.8 billion years ago (Fig. 1.15).

Rosetta found more than craters. Lutetia's face is marked by ridges and faults, traces of a violent past. A complex network of linear structures stretches across long distances of Lutetia's landscape, some up to 80 km in length. Many fractures appear to result from seismic activity that warped some of its craters. The fractures appear to arise from internal forces, rather than as a result of impacts.

Lutetia has an atypical surface composition that does not fit into standard asteroid classifications. Its complex surface may be the remnant of many collisions with asteroids of different materials. Lutetia's odd composition and high density may mean that the asteroid is partially differentiated, with a heavy metallic core surrounded by a primitive crust similar to carbonaceous chondrite asteroids (Fig. 1.16).

Fig. 1.15 An annotated map of the asteroid Lutetia, imaged by ESA's Rosetta comet probe. Features on the asteroid are named after provinces in ancient Rome, including Baetica, Achaia, Etruria, Narbonensis, Noricum, Pannonia and Raetia. Major impact craters are named after cities that existed around the same time as the city of Lutetia, ancient Rome's name for Paris, France. The largest is the 55-km-diameter Massilia (the Roman name for Marseille). Other features have been named after rivers and nearby geography of the ancient Roman empire (Image courtesy of the European Space Agency/Rosetta project)

Fig. 1.16 The diamond-shaped asteroid Steins, imaged by ESA's Rosetta spacecraft, compared to Uluru, or "Ayers Rock" (Asteroid Steins base image © ESA/Rosetta/MPS for OSIRIS Team MPS/UPD/LAM/IAA/SSO/INTA/UPM/DASP/IDA; Uluru photo via Wikipedia Commons; https://commons.wikimedia.org/wiki/File:Uluru_(Helicopter_view)-crop.jpg. Asteroid surface texture added by the author.)

MATHILDE AND 433 EROS

The Rendezvous spacecraft, later christened Shoemaker,[20] embarked on an asteroid orbiting mission in 1996. Built by Johns Hopkins University's Applied Physics Laboratory, NEAR/Shoemaker encountered the asteroid Mathilde on its way to its primary target, 433 Eros. Mathilde was the first C-type asteroid seen at close range. During the 10-km/s flyby, NEAR passed within 12,000 km of the asteroid, carrying out high- resolution imaging and spectral mapping of the mineral-rich asteroid. The best images returned resolutions down to 200 m/pixel. The $66 \times 48 \times 46$-km rock is as dark as asphalt, with a large amount of carbon on its surface. Its density is less than half that of a typical carbonaceous chondrite, perhaps indicating that the asteroid is a very loosely packed pile of rubble. NEAR investigators estimate that up to 50 % of 253 Mathilde's interior is empty space.

Next on the spacecraft's agenda was Eros, the first asteroid to be orbited. NEAR/Shoemaker settled into orbit on February 14, 2000, and finally came to rest on the surface on February 12, 2001. Part of the Amor group of near-Earth asteroids, Eros' orbit carries it from the main belt to just inside the orbit of Mars (Fig. 1.17).

Like 17 % of its siblings, Eros is a stony (S-type) asteroid. It is roughly the same size as the dinosaur-killing Chicxulub impactor, measuring $34.4 \times 11.2 \times 11.2$ km, making it one of only three near-Earth asteroids with diameters above 10 km. The asteroid is elongated, with a dust-enshrouded "saddle" in the middle. Most of the larger boulders scattered across Eros were tossed from a single crater, called Charlois Regio. Compression from the same impact may have created the chasm Hinks Dorsum, on the opposite side from the impact. NEAR/Shoemaker data suggests that Eros could contain 20,000 billion kg of aluminum and similar amounts of "rare-earth" metals such as gold and platinum. Since Eros is irregularly shaped, it has a varying gravity field, making NEAR's orbit tricky. Flight engineers continually corrected its flight path as it circled the irregular space rock (Fig. 1.18).

[20] Named after the pioneering impact geologist Eugene Shoemaker.

Fig. 1.17 Both asteroid 433 Eros (distant view at left and close up at center with Smart Car superimposed for scale) and 253 Mathilde (right) were visited by the NEAR/Shoemaker spacecraft (Image courtesy of NASA/JPL/JHUAPL)

The tiny asteroid Itokawa, just 535 m long and 209 m wide, holds the distinction of being the first asteroid from which samples were returned to Earth. The Japanese Space Agency (JAXA) conceived of the daring mission, which they called Hayabusa ("Falcon"). The Hayabusa spacecraft returned images of the bizarre peanut-shaped Itokawa on its final approach before landing. The little Earth-crossing asteroid appears to be a loosely held together pile of rubble, but the rubble is inconsistent, ranging from sand-sized grains to boulders the size of a ten-story building. Later studies showed that the asteroid spins around a point offset from its center by 21 m. This offset indicates that one lobe of the asteroid is less dense than the other. Researchers[21] now contend that Itokawa is composed of two different materials that gently collided and stuck together to form its current peanut shape. Particles from the Hayabusa craft bear similarity to those of common carbonaceous chondrite meteors. Scientists studying the asteroid material now suggest that the debris making up Itokawa originally formed in the interior of a larger asteroid (Fig. 1.19).

OFF ON A COMET

Comet Hartley 2 (103P/Hartley) and others of its icy kin have been visited only a handful of times. Many comets seem to be peanut- or potato-shaped rather than round. In the case of Comet Hartley, the bowling pin-shaped nucleus is between 1.2 and 1.6 km long. The Deep Impact spacecraft coasted by at an altitude of 700 km in November of 2010. Despite its speed of 44,300 km/h, the craft snapped images of the spectacular active nucleus and sampled its gases. The little comet spins around once each 18 h, but it also spins around a second axis, bobbing and dipping as it turns. Hartley's two tips are rougher than its midsection. They are encrusted with

[21] See "The Internal Structure of Asteroid (25143) Itokawa as revealed by detection of YORP spinup," Lowry, et al., *Astronomy and Astrophysics*, Volume 562, A48.

Fig. 1.19 The asteroid Itokawa, compared to Toronto's CN Tower. In 2005, the Japanese spacecraft Hayabusa landed on the surface, retrieved microscopic particles, and returned them to Earth (CN Tower via Wikimedia Commons by Wladislaw, https://commons.wikimedia.org/wiki/File:Toronto_-_ON_-_CN_Tower_bei_Nacht2.jpg. Asteroid Itokawa courtesy JAXA, Japanese Aerospace eXploration Agency. Art by the author.)

glistening boulders – perhaps chunks of pristine ice – some of them 80-m high (as tall as a sixteen-story building). Frozen carbon dioxide (dry ice) from the interior breaks through the crust, spewing out jets that carry water vapor with them into space. The carbon dioxide is likely preserved from the Solar System's earliest formative years. But at 300 megatons, the comet should survive another 700 years, passing by the Sun a hundred more times.

Hartley is due back to the inner Solar System in the spring of 2017. Like other double-lobed comets (such as Comet 19P/Borrelly, encountered by NASA's Deep Space 1 in 2001 and Rosetta's 67P/Churyumov-Gerasimenko), all have thin necks that may eventually fracture, splitting the comet nucleus into two or more bodies. Often, when this happens the comet is unable to survive, falling apart as it warms near the Sun (Fig. 1.20).

Fig. 1.20 Comet Hartley 2 is roughly six times as tall as the Eiffel Tower. Even a "small" body such as this could wreak havoc on a wilderness area or an urban center, leaving a crater 12–20 km across and a far larger outer swath of destruction (Comet image courtesy of NASA/JPL-Caltech, UMD; Eiffel Tower image courtesy of Alexandra Carroll)

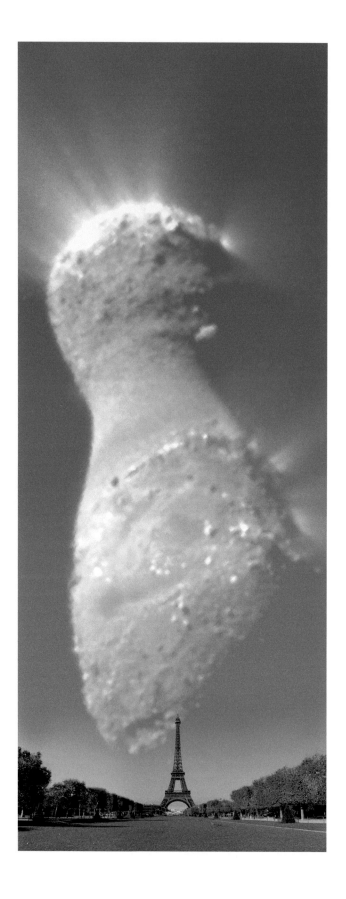

The comets and asteroids survive as remnant brick-and-mortar pieces of our planetary system, time travelers from the earliest formative years of the solar family. And although one occasionally wanders close enough to Earth to make us very nervous, these small bodies offer a valuable window into the history of the planets and their heritage. But before we visit the major worlds, we must understand the ice dwarfs, small moon-worlds in limbo between the asteroid/comets and the chief planets.

Chapter 2
Ice Dwarfs and Tiny Moons

The dwarf planets form a third class of large bodies in the Solar System, behind the mighty gas and ice giant worlds and the smaller terrestrials. One asteroid and several KBOs are considered large enough to fall within this class. To qualify as a dwarf planet, the International Astronomical Union says, the body must directly orbit the Sun (not another planet), it must be massive enough for its gravity to pull it into a sphere, but not large enough to have cleared a pathway in its orbit of the debris that tends to float around throughout the Solar System. The only dwarf planet among the asteroids is Ceres, which was also the first dwarf planet to be explored up close.

CERES

Ceres makes up a third of the mass of the entire Asteroid Belt. Its surface area is equivalent to the subcontinent of India. For over 50 years, it was considered a planet (as were the asteroids Pallas, Juno and Vesta). But in a preview of what was to come with the ice dwarfs and Kuiper Belt, astronomers came to realize that Ceres was part of a larger family of bodies inhabiting an area now known as the Asteroid Belt. Orbiting between Mars and Jupiter in the main Asteroid Belt, Ceres circles the Sun once each 4.6 Earth years. Its day is just over 9 h from sunrise to sunrise.

Ceres orbits between two major regions of our planetary system. Beyond it circle the gas and ice giants, cold worlds whose environs are ruled by ice. Inside of Ceres' neighborhood are the warmer, comparatively dry terrestrial worlds. Astronomers wondered what Ceres, a straddler of worlds, would be like. They had to suffer through many centuries of fuzzy telescopic observing and guesswork, until the arrival of the Dawn spacecraft in 2015. What the spacecraft beheld was an in-between world of rock and ice. The asteroid has a great deal of water-ice, so many researchers predicted that features would be softened because of Ceres' closeness to the Sun. But in fact, the craters and mountains stand out in crisp detail, with well-defined rims, peaks and ridges. "It does come as a surprise to some extent," says Marc Rayman, principle investigator of the Dawn spacecraft now in orbit around Ceres. "But we didn't fully know what to expect, although there were predictions that we would see fewer craters because the relatively warmer ice content of the crust would have caused the surface material to flow and erase most of the craters. But that apparently hasn't happened."

While Ceres is quite different from the more rocky asteroids nearby, evidence suggests it may have formed close to its current distance from the Sun, one of the rare survivors from the earliest epoch of planetary

M. Carroll, *Picture This!: Grasping the Dimensions of Time and Space*, DOI 10.1007/978-3-319-24907-0_2, © Springer International Publishing Switzerland 2016

Fig. 2.1 The famous "Spot 5" outcrops on the floor of Ceres' Occator Crater, first detected by the HST (Image courtesy of NASA/JPL-Caltech/UCLA/ MPS/DLR/IDA)

formation. Another theory proposes that Ceres was originally a Kuiper Belt object that migrated inward, perhaps gravitationally disturbed by a close encounter with another KBO. Yet a third possibility is that the asteroid formed somewhat farther from the Sun than it is now, but not as far as the Kuiper Belt. Models indicate that it is dynamically difficult to get an object from the Kuiper Belt in to the orbit where Ceres resides today. The asteroid's composition will provide insight into its origins, because the mix of different surface materials will reflect the conditions at the location where Ceres formed.

The planetoid appears to be differentiated, with a rocky core and water-ice mantle and crust. The ice layer goes down roughly 100 km before reaching the rocky interior. Before the Dawn mission, estimates indicated that Ceres contained roughly 30 % water-ice by mass, says Rayman, "so there's a lot of rock there."[1]

A variety of materials are scattered across the surface, including rock and ice. Craters tell investigators something about composition as well. On Ceres, their depth and diameter are similar to impact features on Dione and Tethys, two icy satellites of Saturn that are about the same size and density as Ceres. The features on Ceres are consistent with an ice-rich crust.

---

[1] As of this writing, Dawn is still in the process of determining just what that composition is.

Early images from the Dawn spacecraft confirmed several bright spots seen in Hubble Space Telescope images. Preliminary data suggests that the bright sources may be leaking water, underscoring Ceres' soggy disposition. But Marc Rayman warns that his team's close-up readings are not yet in. "Vapor is an unconfirmed observation, and still a matter of internal debate. That's one possible interpretation of the data. It's a very weak signal, so it still needs further scrutiny."

The existence of water vapor at Ceres originally came from data near Earth, from the Herschel Space Observatory, a European Space Agency space telescope. In January of 2014, ESA researchers reported that Herschel had detected an extremely thin veil of water vapor around the asteroid. Due to the limits of resolution, they were unable localize it, but the results constituted a good detection with high confidence. The implication is that water molecules are somehow making their way from Ceres' icy interior out into space. ESA scientists offered two explanations: first, that cryovolcanism was venting water into space through some kind of diffuse clouds or plumes; or second, that ice on the surface was sublimating.

The orbiting telescope did not observe the phenomenon every time it viewed Ceres, so the vapor's sporadic presence is somewhat baffling. The outgassing may be linked to Ceres's distance from the Sun. Ceres has a slightly elliptical (or eccentric) orbit, so there may be just enough difference from perihelion to aphelion to create episodic outgassing. The vapor may also be a result of some other variable geological processes that are occurring on Ceres. Whatever the explanation, the cloud around Ceres was very tenuous. Rayman estimates that the density of water vapor around Ceres during the ESA observations was lower than the atmospheric density a hundred miles above the International Space Station. "In other words, Dawn doesn't need windshield wipers; it's not that dense."

Although the bright regions on Ceres resemble unearthed ice outcrops seen in craters on other moons and Mars, it isn't obvious that Ceres' spots are actually ice. Ceres is likely so warm that ice won't last long on the surface. It will sublimate, melting from an ice directly to a gas. Other possibilities for the spots' composition include salts.

The bright spots depart from the model of excavated ice in another way: their distribution seems to be independent of impact craters. They occur in concentrations. For example, Occator Crater has more than a dozen within it, and yet there are others not related to impact structures at all.

Ceres may have another surprise up its sleeve – an internal ocean buried beneath its ice-rock mantle. Oceans have been confirmed on the moons Europa and Ganymede (at Jupiter) and Enceladus (at Saturn). Ceres' size is somewhere in between Europa and Enceladus, at 940 km. Studies of surface geology will provide clues as to how extensively Ceres has been reshaped by some interaction between surface and subsurface water, past or present.

Fig. 2.2 The dwarf planet Ceres (at center) compared to the moon and earth (Images courtesy of NASA/JPL-Caltech/UCLA/MPS/DLR/IDA)

THE CENTAURS

Beyond the Asteroid Belt, a smattering of planetoids travel around the Sun in circuits from just inside Saturn's orbit out to the realm of Neptune. These are known as the Centaurs. Centaurs constitute a sort of hybrid, sharing characteristics of both comets and asteroids.[2] Their orbits cross (or have crossed in the past) one or more of the giant planets. Some orbits are fairly circular, while others are quite eccentric (out of round). All Centaurs follow unstable orbits that will change dramatically over time as they are perturbed by the major planets. Within a few million years, they will either be shifted into different orbits or tossed out of the Solar System completely. Estimates put the total number of Centaurs larger than 1 km at some 44,000. Although no Centaurs have been encountered up close by spacecraft, some models suggest that Saturn's moon Phoebe is a captured Centaur.

The two largest Centaurs are 2060 Chiron (discovered in 1977) and 10,199 Chariklo (located in 1997). Chiron's orbit carries it just inside the orbit of Saturn and out beyond the orbit of Uranus. Because it exhibits characteristics of a comet, is also has the cometary designation of 95P/Chiron. As it approached its closest point to the Sun in its orbit (perihelion), Chiron developed a coma, the kind of spherical cloud common to comets. Four years later, a comet tail began to stream away from it. The best-measured diameter of the Centaur so far comes in at 233 km, making it a candidate for a dwarf planet.

[2] At least three centaurs exhibit comet-like comas: Chiron, 60,588 Echeclus and 166P/NEAT. The latter, in fact, is classified as a comet.

Chariklo is the largest confirmed Centaur. Its 250-km diameter puts it in the same size class as the mid-sized asteroids. Its path carries it out just far enough to kiss the orbit of Uranus – coming within less than an AU of the green giant, then drops it to within 3 AU outside of Saturn's orbit. And like Saturn, the little object has a system of rings. Its outer ring spans about 3 km wide, while the more extensive inner one covers about 7 km in diameter. The rings are separated by a 9 km gap. Chariklo is the smallest object known to have rings. The rings may explain a spectral mystery. Before 2008, the light coming from Chariklo contained traces of water, but in 2008 those traces disappeared from the spectrum. At the same time that the water signature disappeared from the spectrum, Chariklo's overall brightness dropped. Then, after 2008, Chariklo once again brightened, and water returned to its spectrum. This strange set of events could be explained by a water-ice ring system that tips around the little planetoid. If the rings became edge-on as seen from the Earth in 2008, water would have disappeared from the spectrum, and the bright rings would have all but disappeared. Whatever the case, Chariklo's gravity is so weak, and its rings are so small, that the rings should dissipate within a very few million years. This means that they may be quite young formations. While Chariklo's gravity is not strong enough to hold them for long, if the rings are bracketed by small shepherd moons, as we see at the ring systems of the gas and ice giants (see Chap. 4), the rings could be preserved for much longer. Shepherd moons are typically just a few tens of kilometers across, but even their slight gravity has a profound effect on the rings they accompany. Shepherd moons tug on the particles within a ring, keeping them within a tight boundary. If Chariklo has small moons on either side of its rings, the loops of debris may be preserved from falling apart, kept organized by their watchful shepherds.

THE KUIPER BELT: FIRST DISCOVERIES

Another population of bodies lies out even farther than the asteroids or Centaurs, the aforementioned Kuiper Belt. While Ceres was the first of the main belt asteroids to be discovered, the first Kuiper Belt ice dwarf to be discovered was Pluto. Pluto counts its membership in the great cloud of objects orbiting beyond the realm of Neptune (a few have orbits that carry them just inside of Neptune's track). From 30 AU to about 50 AU from the Sun, some 70,000 icy worldlets are thought to inhabit the Kuiper Belt region. The majority orbit in the same plane as the planets, but the belt spreads out above and below, much like the Asteroid Belt closer in. The Kuiper Belt is 20 times as wide and up to 200 times as massive as the main Asteroid Belt. The inhabitants of this dark dominion represent some of the oldest remnants of the Solar System's early formation. While the Sun burned away many of the volatiles in the territory close to it, leaving the rocky terrestrials, its influence was felt more subtly out where the gas and ice giant planets formed. These worlds contain gases from that primordial time, but their chemistries have changed over time. Not so in the Kuiper Belt, where the dregs of the planetary system remained pristine from the earliest times, changed only slightly by the faint ultraviolet light[3] of the distant Sun.

The poster child of the KBOs, Pluto, holds the distinction of being the only Kuiper Belt object visited up close by a spacecraft.[4] Its orbit around the Sun is so large that the entire history of the United States has passed during just one Plutonian year, or circuit of the Sun.

Why is Pluto no longer a planet? What did the poor little world do to deserve a demotion? Pluto was considered the ninth planet until the discovery, in 1992, of the bizarre object 1992 QB1. This Kuiper Belt object was only the first of many to be discovered in quick succession. Its lazy course around the Sun takes 291 years. Only 167 km across, the object is a third the size of Saturn's small moon Enceladus (see Chap. 3).

With the discovery of so many Pluto-like worlds out there, the International Astronomical Union was faced with a dilemma. Should those small worlds join the list of planets, or did they deserve a class of their own? The IAU finally came to the conclusion that for an object to be a "planet," it must be big enough to be round (a phenomenon known as hydrostatic equilibrium), it must orbit the Sun without orbiting another body, and it must be big enough to clear debris around its orbit. "Dwarf planet" meant that the object was large enough to be in hydrostatic equilibrium and it orbited the Sun, but it was too small to clear debris. To this point, however, the IAU has only a short list of dwarf planets: Ceres, Pluto, Eris, Makemake and Haumea. With the exception of the asteroid Ceres, all of the dwarf planets reside beyond the orbit of Neptune in the Kuiper Belt. Let's take a closer look at a few.

Pluto

The distance to Pluto and the Kuiper Belt defies our normal, day-to-day understanding, but we can try to put it within that mundane context by envisioning a car trip…a long one. A flight to a planet actually carries a spacecraft on a long arc, but with our imaginary auto, we'll travel a straight path. And while Pluto's orbit is not circular, we'll take its average distance of 39 AU. We'll cruise at the nice round speed of 100 km/h. There are no rest stops or gas stations, but there is no gravity either, so we'll travel without stops. Driving constantly at our speed, our little one-way road trip should take us about 6307 years!

Perhaps we need a more efficient mode of transportation. The airspeed of an average Airbus A380 is 900 km/h, and unlike your car, it comes with bars, beauty salons and restaurants. At its formidable speed, we'll be searching for a parking space on Pluto in about 680 years. Fortunately, the aircraft sports a duty-free shop in case we get bored.

[3] Ultraviolet light interacts with methane and other materials, changing their chemistries.

[4] Aside from several short-period comets, including Giacobini-Zinner, Grigg-Skjellerup, Halley, Hartley 2, Borelly, Tempel, Wild and 67P/Churyumov-Gerasimenko.

With its great distance and small size, little was known about the mysterious planet of the underworld for decades after its discovery.

Pluto takes two and a half centuries to coast around the Sun. This leisurely "year" is shorter than some of its siblings, more distant KBOs that may take 550 years to make the circle. The ice dwarf is about half the diameter of the planet Mercury. Its large moon Charon is so big compared to Pluto that the two are considered more of a double planet. Pluto's diameter is 2370 km, plus or minus 3 km. The best Earth telescopes using adaptive optics could discern only general changes in the surface albedo, and the Hubble Space Telescope gave us comparable views, discerning areas several hundred kilometers across. More detailed views would have to await a spacecraft flyby.

Pluto turned out to be a wildly bizarre world in its own right. Our tantalizingly scant knowledge of this distant dwarf planet changed dramatically in July of 2015, when the New Horizons spacecraft coasted through the Pluto system. Built by Johns Hopkins University's Applied Physics Laboratory, the craft made history by visiting the most distant object ever encountered by a spacecraft. Images from its breakneck flight through the Pluto system displayed a heart-shaped region embellishing the globe across one hemisphere at the equator. Tombaugh Regio,[5] named after Pluto's discoverer, spreads across an area the size of the state of Texas. Its left lobe, known as Sputnik Planum, is a flowing ice plain made primarily of frozen nitrogen and carbon monoxide ices. In fact, the ice inventory of Tombaugh Regio may be a reservoir supplying and moderating the thin veneer of atmosphere cocooning Pluto.

[5] As of this writing, names of features on Pluto are provisional, awaiting official acceptance by the International Astronomical Union.

Fig. 2.3 Pluto in false color brings out the subtle differences in the ices and minerals blanketing Pluto's surface. Here, we see the subtle disparity between the left and right lobes of Pluto's "heart," Tombaugh Regio (Image courtesy of NASA/ Johns Hopkins University Applied Physics Laboratory/ Southwest Research Institute)

The behavior of Pluto's atmosphere came as a surprise. In some ways, the little planet is comet-like. As it nears its closest distance to the Sun, it develops an atmosphere, and as it moves back out, its atmosphere collapses onto its surface. Pluto's nearest approach to the Sun occurred in 1989, but in the ensuing years, as the planet began to cool, its atmosphere actually gained in pressure. This baffling trend seems to have broken shortly before New Horizon's arrival. Data indicates that Pluto's atmosphere dropped by half over the 2 years preceding the encounter. Atmospheric experts continue to ponder its mysterious behavior. Haze layers in Pluto's atmosphere extend to at least 120 km, with denser layers at about 37 and 62 km up. Sunlight generates the upper layers of haze when ultraviolet light tears apart the molecules of the methane high in Pluto's atmosphere. The breakdown of methane produces more complex hydrocarbon gases such as ethylene and acetylene. These then combine to form tholins. Tholins are organic materials with a reddish hue, common in the outer Solar System, where methane is common. As these hydrocarbons drift to the lower regions of the atmosphere, they condense into ice particles, forming the hazes. From that haze, a steady snow of organic tholin flakes rain down, collecting in dark areas along Pluto's equator and lending the planet its rich ruddy color.

One thing is clear – the atmosphere frozen to its surface is on the move. At the northern borders of Sputnik Planum, ice flows migrate toward the highlands, surging into the canyons between mountains and

Fig. 2.4 Top: Pluto's unique "Snakeskin terrain" imaged at a range close enough to see objects 1.8 km in size. Bottom: With the Sun behind it and New Horizons bidding it farewell, Pluto displays a halo of atmospheric hazes (Images courtesy of NASA/Johns Hopkins University Applied Physics Laboratory/ Southwest Research Institute)

inundating crater floors. The flow features appear quite similar to terrestrial glaciers, embaying low areas and streaming around obstacles like mountains. Ponds of ice fill some craters, often leaving a central peak sticking out of the middle. Sputnik Planum's alien glaciers are made not of water ice but of frozen methane, nitrogen and carbon monoxide. These bizarre sheets of ice may still be active today. Sputnik's ices appear to be overrunning the landscape on its northern borders. Before our first close-ups of the famous KBO, the only other sites in the Solar System known to host glacial activity were Earth and Mars.

At the western border of Sputnik Planum, the vast plain's nitrogen, methane and carbon monoxide ices thin out, coating the adjacent region with dwindling frosts. Just to the west of Sputnik lies a great dark formation, nicknamed "The Whale." It is the Cthulhu Regio, a sprawling reddish-brown region stretching along 2990 km of Pluto's real estate.

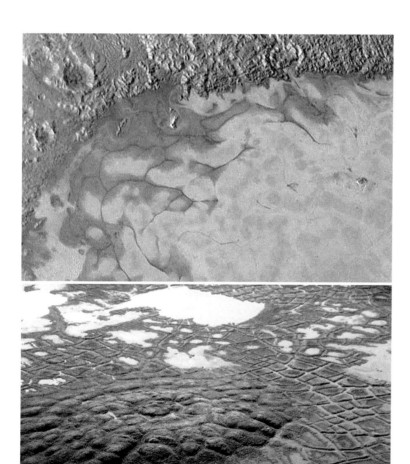

Fig. 2.5 Top: The icy plains of Sputnik Regio flow like glaciers into the mountains to the north (Image courtesy of NASA/Johns Hopkins University Applied Physics Laboratory/Southwest Research Institute.) Bottom: Some ice-related features in Earth's arctic regions resemble the polygonal terrain on Pluto's Sputnik Planun, but the forms probably issue from very different processes (Image by Emma Pike via Wikimedia Commons, https:// commons.wikimedia.org/ wiki/File:Melting_pingo_ wedge_ice.jpg)

The irregular, somewhat spotty formation is the darkest area on the planet. Craters pepper Cthulhu's surface, implying that its surface may be billions of years old. Its ancient terrain provides a dramatic contrast to the fresh, craterless Sputnik Planum next door.

Pluto is a small world with big mountains. Summits puncture through its snowy plains in several major ranges. Within the early views obtained by New Horizons, which only saw one hemisphere in detail, the primary chains seem to huddle fairly near the equator. Spreading out along the southwestern edge of Pluto's Tombaugh Regio, the Hillary mountain chain rears up to a 1.5-km-high altitude, comparable to the Appalachian chain in the eastern United States. But far from the granite of terrestrial peaks, these mountains are made of water, frozen rock-hard at Pluto's chilly temperatures. The Hillary range forms a demarcation between the icy Sputnik Planum and the rugged Cthulhu.

Higher still are the Norgay Montes (Norgay Mountains), named after the Sherpa who, with Edmund Hillary, first summited Mt. Everest. These rugged summits rise up 110 km to the east of the Hillary chain, and reach heights competing with North America's Rocky Mountains. The range marks the southeastern boundary of Tombaugh Regio.

Smaller than the most active moons, many researchers anticipated that the Pluto/Charon system would be geologically quiescent. "We've tended to think of these midsize worlds … as candy-coated lumps of ice," says John Spencer of the Southwest Research Institute, and an investigator on the New Horizons team. But the dramatic landscapes of Pluto and Charon have changed researchers' views of Kuiper Belt objects. "This means they could be equally diverse and be equally amazing, if we ever get a spacecraft out there to see them close up."

Although Pluto's dynamic surface appears to be sculpted by internal forces strong enough to raise chains of mountains, researchers are baffled as to where that energy is coming from. One theory proposes that Pluto's interior energy may issue from the decay of radioactive material in a small, rocky core. But as we have seen, the size of a planetary core matters (see Chap. 1), and radiogenic material was thought to be too scarce within Pluto to have much of an influence on its geology. Another idea is that Pluto's exciting landforms come from energy released by the gradual freezing of an underground ocean.

Whatever the cause, Pluto's varied and lively surface is a game-changer for Kuiper-Belt worlds. Said New Horizons principal investigator Alan Sterns in a recent interview, Pluto has shown us that, "small planets are capable of generating geologic change billions of years after they have formed. That's going to send the geophysicists to the drawing board to figure out how you do that".[6]

[6] *Arizona Daily Star,* Science and Technology Blog, Aug 9 2015.

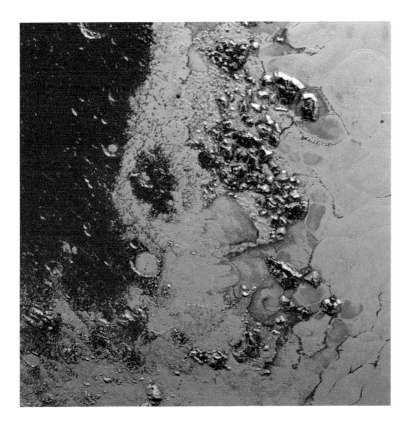

Fig. 2.6 *The spectacular mountains of Pluto are similar in scale to North America's Rocky Mountain range. This view covers an area at the edge of the dark, methane-rich equatorial region (Image courtesy of NASA/Johns Hopkins University Applied Physics Laboratory/Southwest Research Institute)*

Charon

Pluto is tidally locked with its largest moon Charon,[7] keeping the same face toward it. Charon is 1209 km in diameter, about the size of Uranus's moons Ariel and Umbriel.

For a small moon far away from any major planet, Charon has a remarkably fresh face. Its few craters are fairly well preserved, and its surface is bright (though not as bright as Pluto's), implying youthful geology. A series of troughs, including a massive 900-km chasm twice as long as the Grand Canyon, stretch across the equator. The little world wears a darkened polar cap, probably changed in color due to the Sun's radiation of organic ices.

Charon sports a crater-like depression apparently filled with a large object. The 'moated mountain' may be an uplifted mass of material, perhaps cryovolcanic. Much of Charon's surface is blanketed with crystalline ammonia hydrates and water-ice crystals, which do not typically survive in a vacuum for very long. This may mean that Charon's landscape is continually resurfaced by cryovolcanic eruptions. However, some researchers suggest another possibility – Pluto may share its atmosphere with its moon. Leaking atmosphere from Pluto may drift on to the surface of Charon, carrying with it those dark hydrocarbon tholins. There, Charon's lack of gravity, along with the action of sunlight, strips almost all of it away, but at the poles, the colder gases are able to hold on, freezing to the surface. Charon is just coming out of a 60-year-long winter night, so its pole may be chilled enough to hold on to the Plutonian gases.

[7] Other moons include Styx, Nix, Kerberos and Hydra.

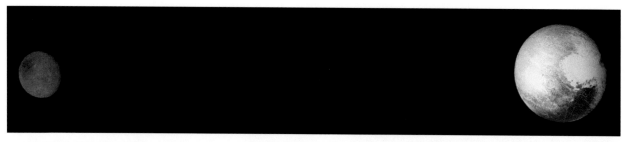

Fig. 2.7 Pluto and Charon, seen in natural color, exhibit striking differences. The two dwarf worlds are seen here at their natural distance from each other. They are tidally locked, always keeping the same face toward each other (Image courtesy of NASA/Johns Hopkins University Applied Physics Laboratory/Southwest Research Institute)

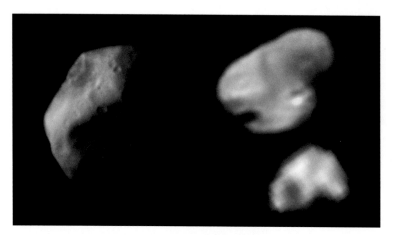

Fig. 2.8 Two small moons of Pluto, Nix (left, in color) and Hydra (Images courtesy of NASA/Johns Hopkins University Applied Physics Laboratory/Southwest Research Institute)

Pluto and Charon orbit around a point between them, qualifying as a double planet. While Charon and Pluto essentially orbit each other, Charon and four smaller moons orbit in synch with each other. This suggests that the clan formed from a debris cloud, perhaps the result of a cosmic crackup. Charon has more ice and less rock than Pluto, implying that it consists of material similar to that of Pluto's outer layers. Earth's own Moon arose from an impact on Earth. A glancing blow from a Mars-sized object peeled off outer layers of our young planet. Those layers condensed into the Moon. A similar scenario may have occurred at Pluto, resulting in a cloud of ice and dust that eventually condensed into Pluto's system of moons.

Pluto's other moons, in increasing distance from the dwarf planet, are Styx, Nix, Kerberous and Hydra. One moon seems set apart from the others: Kerberos. That moon is much darker than the others, so some planetary scientists conjecture that Kerberos is a remnant of the original impactor, while the other satellites consist of the material peeled off from Pluto itself.

The smaller ice moon Nix spans just 35 km, while Hydra is 45 km across. They are comparable to Saturn's moons Helene and Atlas. Kerberos and Styx are smaller still.

Other Kuiper Belt Objects

If Pluto has taught us one thing, it is that the Kuiper Belt objects – far from being dead, frozen, flying rocks – may be quite diverse in nature. A brief survey of the family of objects in the Kuiper Belt underscores the point. Shortly after the discovery of QB1, researchers discovered the much larger Quaoar, larger than Ceres but smaller than Pluto. The mysterious ice world was discovered by the Hubble Space Telescope, which estimated the object's

Fig. 2.9 The dwarf planets compared to the Iberian peninsula of Spain and Portugal. Left to right: Ceres, Pluto, Haumea, Makemake, and Eris. Note that surface details of Haumea, Makemake and Eris are artistic approximations (Ceres courtesy NASA/JPL/Dawn; Pluto courtesy JHUAPL/New Horizons; art by the author)

diameter at 1092 km. Quaoar is a planetoid the discovery of which helped the IAU to create its new classification of "dwarf planets" and the redefining of Pluto as an "ice dwarf planet." Quaoar may be massive enough to be considered a dwarf planet, but due to uncertainties of its size and mass, it has not been classified as such yet. Quaoar lurks at about 42 AU from the Sun, a billion kilometers farther out than Neptune. It takes 288 years for Quaoar to journey once around the Sun.

The dwarf planet Haumea follows an orbit ranging from 35 to 51 AU. At this great distance, the icy world takes 284 years to make one circuit of the Sun. Haumea is one of the more oddball KBOs yet found. With a mass a third that of Pluto, the little world is plenty large enough to have relaxed into a sphere due to its own gravity. But it hasn't. Haumea is decidedly egg-shaped, with one axis twice as long as the other. The dwarf planet is roughly 2000 × 1500 × 1000 km. Its rapid rotation provides a day just 4 h long. At least two moons circle Haumea.

"Haumea is without a doubt the weirdest object out there," says Caltech planet-hunter Mike Brown, who discovered Haumea in 2004. "It's the fastest spinning large object that we know of in the Solar System. Just those facts make it exotic, but it's also got two moons and a whole family of KBOs leading it and trailing it." Astronomers now understand what's going on with Haumea's strange family of moons and ice fragments. The little football-shaped world used to be a much larger object, but another wanderer smacked into it early in the history of our planetary system. This glancing blow got Haumea spinning rapidly. That spin pulled it out to the shape it is and partially broke it apart, Brown explains. "Those chunks went flying off into space. We've been tracking those. Those chunks are distributed completely around the Solar System in random places along [Haumea's] orbit."

Haumea's surface is covered in water-ice, making it as bright as newly fallen snow. It appears to have a dramatic reddish spot on one hemisphere, perhaps akin to the spot on Pluto's Nix, but the region's nature is as yet unknown.

Makemake is two-thirds the diameter of Pluto, and reaches 53 AU at its farthest from the Sun. Like most Kuiper Belt objects, its 310-year orbit is inclined to that of the Solar System's major planets. It is the third largest KBO, up to 1502 km in diameter. The extremely cold surface (−243 °C) is probably blanketed by frozen methane, ethane and possibly nitrogen. Makemake is the mythological creator of humanity and god of fertility in the pantheon of the Rapanui people of Easter Island. It has a reddish hue, and methane has been detected strongly on its surface. The presence of methane and possibly nitrogen insinuates that Makemake could have a transient atmosphere similar to that of Pluto's.

Eris was thought to be the ninth-largest object orbiting the Sun, slightly larger than Pluto. But New Horizons spacecraft data tell us that Pluto is a bit larger, probably by a scant 30 or 40 km. Eris likely spans some 2336 km across (with an uncertainty of ±12 km),[8] and weighs in as a quarter more massive than Pluto. Eris was the Greek goddess of strife and discord. Her 557 year-long orbit is certainly discordant, wandering from 38 to 98 AU. About 800 years from now, Eris will actually be closer to the Sun than Pluto ever gets. The dwarf planet approaches close enough to the Sun for surface ices to sublimate, floating away in its low gravity. In this way, Eris resembles a huge comet nucleus. The presence of methane suggests either that Eris has always resided far enough from the Sun for methane ice to persist, or that the frozen worldlet has an internal source of methane that is replenishing gas that escapes from its atmosphere.

Eris has a small moon, Dysnomia, which circles every 16 days. In mythology, Dysnomia was Eris' daughter, the goddess of lawlessness. Estimates of its diameter put the chilly moon at between 250 and 490 km, on a par with Saturn's moon Mimas.

Sedna is the farthest dwarf planet known in our Solar System. It ranges up to 31 times as far from the Sun as Neptune. Sedna is important because in understanding its unusual orbit, researchers will likely gain insight into the origin and early evolution of the Solar System. Sedna may be part of the "scattered disk," a group of objects sent into highly elongated orbits by the gravitational influence of Neptune. But it is so far out that some want to classify it in a separate category called the extended scattered disk. Objects in this Stygian wilderness are too dim to be spotted except when approaching perihelion in the inner Solar System. Sedna's diameter is estimated at between 915 and 1075 km, making it smaller than Pluto's moon Charon, perhaps about the size of Saturn's Tethys. Sedna's spectrum is similar to that of Neptune's geyser-ridden moon Triton. It has the longest orbital period of any known large object in the Solar System, lasting 11,400 years. Only some long-period comets orbit farther than Sedna.

[8] Since Eris has no atmosphere, at least at the moment, detailed measurements have been possible by using stellar occultations. In 2010, observers watched Eris cross in front a dim star. Since they had already charted Eris' orbit accurately, they knew how fast it was moving. They simply timed the star's winking out and reappearance to determine Eris's size. Observations were recorded from multiple locations on Earth, so that a map of the Kuiper Belt object's outer edge could be assembled. Eris turned out to be smaller than initial approximations; its bright, reflective surface threw off astronomers' estimates.

Fig. 2.10 Eris (back), Ceres and Pluto's moon Charon compared to the islands of Japan (NASA photo of Japan; Charon courtesy of JHUAPL/ New Horizons. Ceres image courtesy of NASA/JPL-Caltech/UCLA/MPS/DLR/IDA; art by the author)

Other Kuiper worlds not yet given official planetary status include Ixion and Varuna, both of which measure roughly 500 km across. Ixion is the larger of the two, and may measure as large as 650 km (slightly larger than Saturn's Enceladus). The difficulty in discerning a distant object's diameter is that the surface brightness isn't known with certainty. If an object is dark, it will be larger than if it is bright, because astronomers tally the total amount of light coming from an object to estimate its size. In Ixion's case, the little world is reddish, a common color in the outer Solar System, especially for objects that have methane on their surfaces. Its surface is also covered in carbon, clays, water-ice and organic compounds. Ixion's orbit is elliptical. It is now about 41 AU from the Sun, but it is possible that it will develop a cloud-like coma as it reaches its nearest approach to the Sun at a distance of 30 AU. It will reach this spot, known as perihelion, in 2070. Its distant orbit takes the dwarf world around the Sun just once each 250 years. Ixion's orbit is tilted similarly to that of Pluto, but it is nearly circular, extending out to about 43 AU.

The icy dwarf Varuna is a trans-Neptunian object (its pathway crosses the orbit of Neptune). Similar to Quaoar in its orbit, Varuna takes 283 years to make the circuit of the Sun. Varuna's day lasts 6 h and 20 min. Because of its rapid spin and strange light curve, astronomers believe that Varuna is egg-shaped. The small iceball's true size is a bit of a puzzle. The scale of Kuiper Belt objects can be discerned by a combination of their reflected light and their temperature. If these two values can be charted accurately, their size can be computed. But the distant objects of the Kuiper Belt have low temperatures that are difficult to read. These uncertainties result in a range of estimates for Varuna, from 500 to 1060 km. A recent study using the Herschel Space Observatory – a spacecraft equipped to see heat signatures – suggests the Varuna is 668 km across, with uncertainties ranging from 154 km larger to 86 km smaller. Its shape may be similar to Saturn's Hyperion, but it may be twice as big. Like Quaoar and Ixion, Varuna's surface is reddish.

Another candidate for dwarf planethood is called 90,482 Orcas. Orcas is now some 48 AU from the Sun, on its way out to its farthest distance, which it will reach in the year 2019. It will swoop to within 27.8 AU when it reaches perihelion in 123 years. Its peculiar orbit is in resonance with Neptune, as Pluto's is. For each two orbits that Orcas makes around the Sun, Neptune takes three. But the gravitational dance between Neptune, Pluto and Orcas guarantees that Orcas will always be on the opposite side of its orbit from Pluto. Because of this, Orcas has been called the "anti-Pluto." Studies of its light indicate that water on its surface is frozen into fluffy crystals, which may be a sign of past cryovolcanic activity. It is roughly 800 km wide (a little smaller than Saturn's moon Tethys). Aside from distance, Orcas has something else in common with Pluto. It has at least one moon, called Vanth. Vanth circles Orcas every 9.7 days. If it is tidally locked, as most moons are, this would mean that Orcas has a day identical to Vanth's orbital period. Vanth itself is a bit of a mystery. Its diameter is anywhere from 280 to 380 km. Its color is quite different from Orcas, so it may be a captured Kuiper Belt object.

Two other distant objects have warranted names. They are 120,347 Salacia, at an estimated diameter of 850 km, and 19,521 Chaos. Salacia[9] circles the Sun at an average distance slightly farther out than Pluto. Its spectrum shows less than 5 % water, making it a rocky body. It is the darkest of any Trans-Neptunian objects (Neptune crossers) yet found. Salacia has a small moon called Actaea. The little moon's estimated size is 286 ± 24 km, similar to Jupiter's Amalthea.

For its part, Chaos measures roughly 600 km in diameter, with uncertainty of 140 km on either side. Like other Trans-Neptunian objects, Chaos is part of a family known as cubewanos (named after the first object of its kind discovered, labeled QB1-o). Cubewanos do not cross the orbit of Neptune but remain outside of it, ranging from 40 to 50 AU average distances. They are not in resonance with Neptune (or any other planet).

Other objects that may 1 day be granted dwarf planet status range from 599 to 724 km. All of their orbits lie within the Kuiper Belt, out as far as 50 AU. This distance is significant, says astronomer David Jewitt. "Inside of 50 AU or so, we probably know of all the big guys. When you go further out, you quickly come to a wall. It's the same wall that prevented us from seeing the Kuiper Belt forever and a day. It basically becomes very hard to see objects, even if they're big."

Beyond 50 AU, many large objects may lie undetected. Adding to the difficulty is that many of the Kuiper Belt members follow eccentric orbits. Some might wander in close to Neptune, briefly, but spend most of their lives much further away, so that they escape detection. What might still be lurking out there? In fact planets the size of Earths and Jupiters might still lurk in the outer darkness, far enough away to escape detection. There's plenty of unexplored space out there, because researchers are looking at reflected light. Light fades as an inverse square on the way out, and an inverse square again coming back again. This means that an object ten times farther away is 10,000 times fainter. That's a daunting piece of physics to try to overcome. An Earth-sized planet 620 AU from the Sun would have escaped any survey done to date. Even mighty Jupiter would disappear from current instruments at a distance of 2140 AU, which is still well within the gravitational control of the Sun. "The bottom line," declares Jewitt, "is that the outer Solar System is not known."

To date, one astronomer has found more dwarf planets than any other observer. Mike Brown, professor of Planetary Astronomy at the California Institute of Technology, discovered Eris, which was the first accepted dwarf planet since the finding of Pluto. More would follow in quick succession, and by 2005, a parade of distant objects, many still poorly understood, joined the ranks as possible dwarf planets: Quoaor, Sedna, Makemake and Haumea.

The search for distant worlds was, at times, tedious and time-consuming, Brown remembers. "It's a long-term process. A lot of the hard work is in setting things up for everything to work right and keeping it running night after night after night. Then in the day I would look at the data from the night before. Every time you discover something new, you just get a little emotional charge; you're seeing an object going around the Sun that no human being has ever seen before. There's just nothing better. There was a stage in 2005–2006 when we were really going strong, finding these big ones and these bright ones, and I remember it as the most fun period in my life."

Before the discovery of Eris and several other dwarfs, our view of the Solar System was simple and organized, Brown explains. "We had eight nice neat planets and a ninth oddball that didn't make any sense. It was a lot smaller than anything else and had a weird orbit that was elongated and tilted. For many years Pluto was just a really strange thing that didn't fit in with the rest of the planets. Now, it fits in perfectly and makes sense. We have four major components of the Solar System. We have these medium-sized terrestrial things, and we've got gas giants on the outside and asteroids in the middle, where things were too crazy for any planets to form, and this straggle of Kuiper Belt objects on the outside where things were also just a little too rough to form a planet. So you either have planets, which are the big bullies of the neighborhood, in their circular orbits because nobody can mess with them, or you have these other objects, which always come too close to planets, get knocked around and kicked on to eccentric inclined orbits. You can look at the planetary system and pick out the difference between the big planets and everything else just by looking at where they go."

As Jan Oort surmised, beyond the Kuiper Belt lies the Oort Cloud, home to the most distant objects orbiting the Sun, and residence of the long-period comets (comets with orbits lasting more than 200 years). The cloud of frozen debris and drifting ice balls scatters out to 50,000 AU or more from the Sun. That's eight tenths of a light year.[10] The Oort Cloud forms a globe surrounding the fringes of our Sun's family. The comets and ice dwarf planets in the cloud may equal 100 Earths in mass. We cannot see the Oort Cloud directly. We infer its existence from objects that appear to be deflected from it and which enter the planetary region, but we cannot see objects out that far. They are simply too faint.

Despite its remote locale, the Oort Cloud forms a significant part of our Solar System. The population of bodies in the Kuiper Belt is a thousand times the population in the Asteroid Belt, and the number of bodies within the Oort cloud is vastly larger still.

As we have seen, material of the Oort Cloud's primordial detritus may have been scattered into the outer depths at the hands of the migrating gas and ice giants. The gravity of this massive quartet could have tossed the Oort's theorized dwarf planets and iceballs into the extreme orbits we see now. But some research suggests that a large number of Oort Cloud dwarfs came from the Sun's nearby sibling stars as they formed close in and drifted away. Early in our Solar System's history, other stars were forming in close proximity to the Sun. They were also birthing their own planetary systems, some of which might have exchanged members with the Sun's own. It is possible that many Oort Cloud objects did not form near our Sun but came in from the cold.

Distance and faint sunlight make it difficult to find objects as remote as Oort Cloud members. But members of this distant realm sometimes visit us. Long-period comets, some following paths that take 100,000 years to complete, wander into the inner Solar System. They are far rarer than those dropping by from the Kuiper Belt, but because they have seldom been warmed by the Sun, their fresh ices often create spectacular cometary tails. The two most recent apparitions of Oort Cloud visitors were the comets Hyakutake in 1996 and Hale-Bopp in 1997. Hyakutake barreled into the inner Solar System on an orbit lasting for 17,000 years, but after its interaction with the gravity of the Sun and planets, its orbit will not bring it back for another 70,000. Its tail stretched at least 500 million km. Hale-Bopp made its appearance the next year in grand style. It came in nearly perpendicular to the disk of our planets (the ecliptic). Its last visit was 4200 years ago (2215 B.C.), but this trip, the comet's orbit was trimmed because of a close pass to Jupiter. It will next arrive a scant two millennia from now. The comet is large, with a nucleus six times as far across as Comet Halley's. This is undoubtedly because it has seldom been warmed by the Sun.

Just how far out is the far edge of the cloud? The vast distances between our inner planetary family and the far edge of the Oort Cloud are tough to fathom, but a tour of the world's tallest building may help. Dubai's Burj Khalifa rises 838 m above the Arabian Peninsula. It boasts the highest observation deck in the world, and houses five-star hotels and residential neighborhoods. Using the Burj as a yardstick, we shrink the Solar System down and plop it on the ground floor of the concourse level, the base of the building. For our model, we'll turn the Solar System on its side, with north

[10] Some estimates put this number as high as 100,000 AU, or 1.87 light-years.

and south parallel to the floor. Our central Solar System, with all the major planets and moons and asteroids, has shrunken to the size in which one astronomical unit equals 1 cm. At this scale, with our Sun on the floor, Earth is just 1 cm above. Neptune is 30 cm up, while Pluto, at its farthest, rests just at the meter line, waist-level in the room. The Kuiper Belt extends up another 20 cm, reaching the height of an average adult woman. This doughnut of comets and rubble is long faded away as we reach the ceiling.

The most distant known large object orbiting the Sun, Sedna, lies in the dim reaches of space, but here it floats halfway between floor and ceiling of the third story. The Burj still has 161 floors overhead.

As we ascend past the tenth floor, we enter the inner edge of the Oort Cloud. The Sun is a bright star on the decorative tile some 37 m beneath us. On the 38th floor, we reach the plush Armani Hotel, with gardens, observation areas, lounges and swimming pools. The highest swimming pool in the world greets us at floor 76, but we are now just halfway through our cosmic architectural journey. Any comet falling sunward from this height would take twenty centuries to arrive at the Sun.

We reach 103 stories, 380 m, and pass the height of the Empire State Building. A few floors up, as we cross the 414-m mark, we note the top of another tall building in the Dubai cityscape outside. It is the Princess Tower, second-tallest residential skyscraper in the world.

We continue to ascend, imagining our journey through the dark emptiness of space. Now the buildings around us begin to fall away. Our glass elevator drifts through layers of mist, and soars above flying birds and low-flying private airplanes. On level 123 we spot the Sky Lobby business lounge and library, and on the floor above it, the world's highest observation deck, called "At the Top." From here, we can actually see the country of Iran. But the Oort Cloud is still with us. Its frigid chunks of primordial ice are thinning out, but the edge of the cloud hovers somewhere above.

Music builds as our elevator reaches the 143rd floor, home of the world's highest nightclub. The building is 10 °C cooler here than it was at ground level. It's a good place to stop and party. This level marks the farthest edge of the Oort Cloud, and the approximate edge of our Solar System's influence. Our scale model has brought us to an altitude of 58,500 cm (585 m). At this distance, the Sun is just a pinpoint of light among many. Some estimates put the Oort Cloud at nearly twice this distance.

Mike Brown's running count of dwarf planets, collated last in November of 2013, stands at 10 objects that are "nearly certainly" dwarf planets, 23 of which are "highly likely" to be dwarf planets, 49 "likely," and 86 probable dwarfs.

The size of cosmic objects has an important bearing on their very natures, Brown suggests. "Mass is everything. The very small mass objects are weird and irregular and loosely held together. Think of the comet [Rosetta is] orbiting right now that looks like a rubber duck.[11] Small, low mass things are random conglomerates. As you add mass, the object gets big enough that it can compact itself into a roundish shape. As you add

[11] Comet 67P/Churyumov-Gerasimenko, orbited by ESA's Rosetta spacecraft.

Fig. 2.11 The Burj Khalifa,
tallest building in the world,
serves as a vertical model for
our Solar System out to the
farthest fringe of the Oort
Cloud (Art by the author,
digitally remastered photo
from personal collection.)

Outer Oort
Cloud:
~500 meters/
floor 143

Level 124:
World's highest
observation
deck

highest
swimming
pool, floor 76

Sedna
2½ floors up

Plutinos/
edge of
Kuiper belt
lower 2nd floor

Pluto/
main planets:
first floor

even more, you're compressing the interior, which means you're doing geology, you're heating the interior and having things flow onto the surface, if you're icy or rocky. If you heat things enough, you get things like Earth with plate tectonics and volcanoes and all kinds of crazy stuff. As far as we know, adding more mass doesn't make you a bigger Earth. It makes you into a gassy planet. First, you get the ice giants like Uranus and Neptune, which are these icy cores surrounded by gassy envelopes. Then you add even more mass and you get these super massive things like Jupiter and Saturn, which are gas giants."

SMALL MOONS AND ASTEROIDS

Many of the objects in the Kuiper Belt and Oort Cloud are small, and some may have wandered in from the cosmic hinterlands to be captured by the major planets themselves, becoming moons. The gas and ice giants hold on to extended families of tiny moons. The inner moons, starting at just 181,300 km away, follow orbits that are flattened to the plain of the equator.[12] But the more distant small moons orbit in unusual circuits that can be quite inclined to the equator and extremely eccentric (non-circular). Jupiter alone has 67 confirmed moons. Its four Galilean satellites are large, with Ganymede measuring larger than the planet Mercury. The 63 others are quite small by comparison. The largest of the minor moons is potato-shaped Amalthea, 250 km at its longest span, and 128 km at its most narrow. Amalthea is likely a captured object from the Asteroid Belt. It is so close to Jupiter that it orbits in the equatorial plane, its path made regular by Jupiter's mighty gravity.

From the various moons of Jupiter, the distance to the colossal planet would be obvious just by looking at the apparent size of Jupiter in the sky. For example, from close-in Amalthea, a full Jupiter would appear as an orb 93 times as far across as Earth's full Moon in our sky. From volcanic Io, Jupiter would appear as far across as 39 full Moons. From Europa, 671,000 km out, Jupiter subtends 12.2°, equal to 25 full Moons. Ganymede's Jupiter, over a million kilometers away, is 15 full Moons across, while remote Callisto sees the king of worlds as 9 full moons in diameter, still quite dramatic in its velvet black sky.

Farther out, the moons begin to travel in less circular paths. Typical of these is tiny Himalia, just 170 km across. Himalia is the largest of Jupiter's irregular satellites. It is irregular in shape, and also irregular in the way it orbits the giant world. Its 250-day oblong path takes it to within 9,783,000 km of Jupiter, while swinging out as far as 13,000,000. And, unlike the equatorial-orbiting Galilean moons, Himalia's orbit is inclined to Jupiter's equator by some 29.6°.

Of Jupiter's 67 total moons, the rest follow fairly circular, pro-grade orbits, ranging in size from planet-sized Ganymede to moons such as S/2003 J9, so small (~1 km) that they have yet to be named. Many of the small moons orbit at extreme distances.

[12] At least in cosmological terms: tens or hundreds of millions rather than billions of years.

Fig. 2.12 The co-orbital moons Epimetheus (left) and Janus compared to the island of Crete. Although the little moons are similar in size, notice how dramatically their surface textures differ (Crete photo courtesy of NASA. Moon photos courtesy of NASA/JPL/SSI.)

In fact, we find small moons in odd orbits at great distances from all the giant planets. Two of Saturn's small moons have the unique distinction of sharing the same orbit. Epimetheus and Janus circle the Lord of the Rings at a distance of just over 151,000 km, about half the distance from Earth to the Moon. Both are irregularly shaped. Janus is 179 km across, while Epimetheus is about 116 km in diameter. Their shared orbit is a dance that shows off the way gravity works.

The farther out from a planet an object is, the more slowly it revolves around the planet. The inner of these co-orbiting moons circles Saturn more quickly, but only slightly, completing the course 30 s faster than its sibling. In the course of each Epimetheus/Janus day, the outer moon lags behind the inner one by an additional quarter of a degree. Eventually, the inner moon catches up to the outer one. The gravity of the outer moon pulls upon the inner one, causing the inner moon to speed up and move into a higher orbit. The inner moon, in turn, slows the outer one, dragging it into a lower orbit. The moons trade places in their orbits once every 4 years. Janus's speed is affected less, because it is four times as massive as Epimetheus. The two little moons never approach closer than 10,000 km. At closest approach, they appear as star-like objects to each other. Their last exchange took place in 2014, when Janus's orbital path dropped down toward Saturn by about 20 km, while the track of Epimetheus rose by about 80 km.[13]

Several other moons fall into a similar size range to these co-orbital siblings. Pandora clocks in with a 64-km midsection, but is extremely elongated, with its other axis measuring 104 km. Pandora orbits just outside of Saturn's F-ring, while a smaller moon, Prometheus, orbits inside of it. Fine, talcum powder-sized ice dusts Pandora's surface. Its craters are filled with

[13] The reason for the difference is that Epimetheus is one fourth as massive as Janus.

debris, and fractures beneath the surface may contribute to ridges and grooves on the little moon. Rather than being a great boulder, Pandora may be a loosely bound pile of ice rubble. For its part, Prometheus is even smoother, perhaps indicating a deeper blanket of dust. Prometheus is a shepherd moon of the F ring, herding the fine ring's particles into a thin line around Saturn.

Each 6.2 years, the orbital tracks of Prometheus and Pandora bring them to within 1400 km of each other. During this close approach, the moons would see each other in their respective skies as 6–11 times the size of a full Moon in Earth's sky.[14]

Helene is a co-orbital moon of a different stripe. It hovers in a stable spot 60° ahead of the major moon Dione, but in the same orbit. Tiny Polydeuces trails Dione by the same distance. Helene measures roughly 36 by 32 by 30 km.[15] A series of gullies smear its face, perhaps consisting of powder or dust. The channels resemble water gullies on Mars or Earth. But Helene operates in a vacuum, being far too small to have an atmosphere or liquid water. Instead, the gullies comprise slopes where streams of the fine material flow toward the lowest local spots. Ridges of solid material project from the dusty landscape, marking crater rims and other high points.

Some moons are actually embedded within Saturn's ring system, circling the golden globe in the midst of a blizzard of ice chunks and powder. Their curious location leads them to have an even more curious shape. The moons Pan and Atlas each sport a remarkable bulge around their equator.

[14] From Prometheus, Pandora would appear as a 3.25° object in its sky. From Pandora, larger Prometheus would subtend an angle of 5.5° (compared to Earth's Moon, which is just 0.5° in our sky).

[15] It is similar in size to Mars's largest moon, Phobos.

Fig. 2.13 An assortment of the Solar System's smaller moons, not to scale. Top left to right: Saturn moons Helene (40 km), Pandora (81 km), and Calypso (20 km), top and side views of flying-saucer-shaped moon Atlas (32 km), and three views of Jupiter's Amalthea. Bottom: Two views of Prometheus (Amalthea images from NASA/JPL Galileo project; all other images courtesy NASA/JPL/SSI)

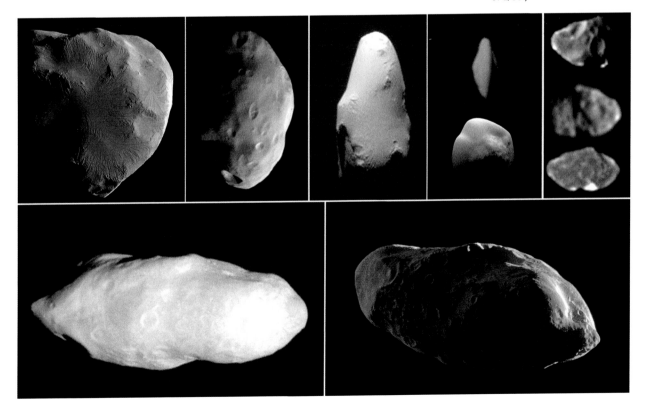

Fig. 2.14 *The miniscule moon Methone is as smooth as an egg. Here, we see the nearly featureless satellite with Dubai's Burj Khalifa, world's tallest building, on one end (Methone image courtesy of NASA/JPL/SSI)*

The ridges consist of ring material that has settled around the middle, giving both satellites a flying saucer appearance. Atlas's rim rises as high off the moon as 3 km, as high as Greece's Mount Olympus. Atlas itself is 39 km across at its widest, but from pole to pole it is only 18 km. Its ridge is lopsided, more rounded (less like a disc) where it faces toward Saturn and also in the direction it travels around the planet. Pan's belt rises some 4 km. This is remarkable, considering that the moon itself is only 33 km across at its widest.

Dozens more small moons orbit the ringed world outside of the spectacular rings, at greater distances from the planet. One of the strangest is Methone. Scientists expected the tiny moon, just 3.2 km across, to be a battered stone, but its surface is blanketed by some kind of dust or powder that seems to mask all topography, sculpting the little moon into a smooth egg. Its leading hemisphere is dramatically darker than the surrounding terrain.

A surprising feature of many small moons was the presence of flows. In the practically non-existent gravity of these small bodies, researchers expected surface materials to rest in place, but many small moons, comets and asteroids bear the scars of migrating matter and avalanches. The Saturnian moons Helene and Calypso are the most dramatic examples, but landslides have also been seen on the asteroids Eros, Vesta and Lutetia, and on several comets, including P/67 and Temple 1.

Landslides, flows and even tracks of tumbled boulders have been spotted on low-gravity moons. These events must have taken place in slow motion over long periods of time. Everything in a low gravity environment moves slowly, as we saw in transmissions of the Apollo astronauts working in the one-sixth gravity of the lunar surface.

To get an idea of how disorienting a low gravity environment such as this can be, we look to a magnificent image of comet 67P/Churyumov-Gerasimenko, taken by the European Space Agency's Rosetta spacecraft. The P/67's features seem, at first glance, quite terrestrial. Mountains rise up in alpine grandeur, with a talus of debris resting between peaks. The Princess Cruise Line's flagship *Golden Princess* is added to the scene for scale. But this vista is deceiving. With such low gravity, it is just as likely that we would experience this region standing on the sandy saddle between mountains, and in the low gravity this plain would become our "ground." The sense of gravity would be so subtle that we would scarcely feel it, and our perception of its direction would be equally nuanced. We have turned Rosetta's scene to illustrate the effect. In doing so, we see how skewed the rock formations can be in the microgravity environment.

With the completion of our mini-moon tour, we now increase our scale, visiting some of the larger, and weirder, moons of the Solar System.

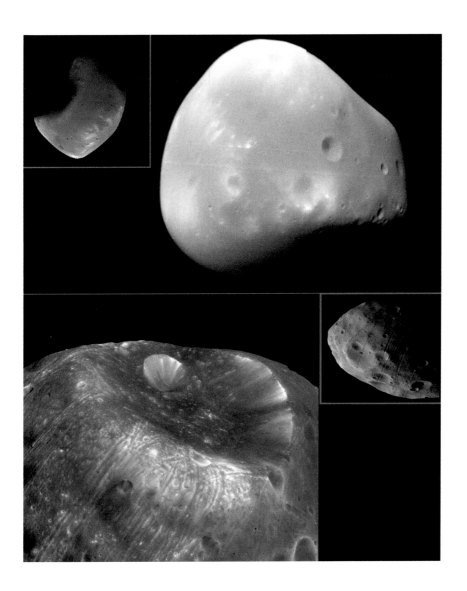

Fig. 2.15 Views of Mars'
irregular, potato-shaped
moons Deimos (top) and
Phobos (Courtesy NASA/JPL/
Malin Space Science Systems)

Fig. 2.16 Two views of the
comet 67P/Churyumov-
Gerasimenko, a rugged,
surreal landscape (Comet
images courtesy of ESA.
Golden Princess courtesy of
Princess Cruise Lines)

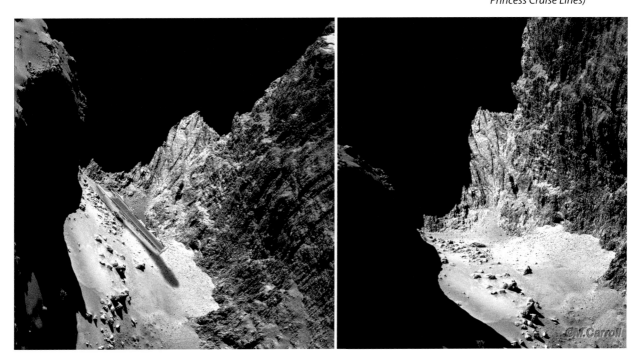

Chapter 3
The Weirdest Moons, Large and Small

Several factors profoundly affect the nature of planets and moons. The first, naturally, is size. Small objects such as asteroids or comet nuclei retain irregular shapes, while larger bodies such as planets and stars become spherical. The shape of an object is not only tied into its size but also its gravity. The larger something is, the more mass it has. The more mass an object possesses, the more gravity it will have. Gravity pulls toward the center of mass of an object. A planet is round because all of its material is being pulled toward its center; a sphere is nature's most efficient shape. We will see this fact played out from scale to scale as we venture from the smallest of rounded moons to the great structures of entire galaxies, huge islands of stars.

A planet's composition also determines its nature. Two worlds identical in size will have different masses – and gravity – if one is made of ice and the other rock. For exam-

Fig. 3.1 *The Galilean satellites compared to Earth and its Moon.* From upper right: *Earth's Moon, Io, Europa, Ganymede, and Callisto* (Images NASA/JPL)

ple, Jupiter's moon Ganymede is nearly identical in size to the planet Mercury. But Mercury has an iron core and stony crust, while Ganymede's smaller rock core is entombed in a deep shell of water-ice, far less dense than rock. Ganymede's gravity measures 1.4 m/s (about 14 % of Earth's, which is 9.8 m/s), while Mercury's is a much stronger 3.7 m/s.

Life in differing gravity environments can present strange effects. On Earth's Moon, gravity is a sixth as strong as on terra firma. If you drop your geology hammer, you have six times as long to catch it as you do on Earth. Structures can be built differently for different environments. Buildings on the small moons of outer planets can be built with lighter materials. Structures in low gravity become taller and more irregular, as we will see among the asteroids and comets.

M. Carroll, *Picture This!: Grasping the Dimensions of Time and Space*,
DOI 10.1007/978-3-319-24907-0_3, © Springer International Publishing Switzerland 2016

Fig. 3.2 Greece's architectural wonder, the Parthenon, would take on dramatically different dimensions in the lower gravity of Mars
(Photo courtesy of Kirstin Canaday, used with permission)

There is a rule of thumb in planetary geology associated with size: the larger a planet or moon, the more likely it is to be geologically active. Larger planetary cores can hold on to the heat of creation longer, and that heat comes out as changes in geological activity such as volcanoes and mountain building. Earth and Venus, the two largest terrestrial planets, are also the most geologically active. Mars comes in next, not only in size but also in fresh geological features. Mercury and the Moon are smallest, hosting surfaces that are – for the most part – geologically quiet, although both have seen volcanism in their pasts. The two asteroid-sized moons of Mars, Phobos and Deimos, are inert, cratered worlds. Large is geologically active; small is quiet and dead. But the moons of the outer Solar System regularly belie this rule.

IO

The first hint that some geological rules were being broken came from Jupiter's moon Io. Io makes up part of a famous foursome, the Galilean satellites (named after their discoverer, Galileo Galilei, using one of his earliest telescope designs). Before the Space Age, astronomers knew that Io exhibited some bizarre behaviors. Its color departed from the color of its siblings, and it seemed to be cocooned by a cloud of sodium vapor. At times, the little world would brighten dramatically. But Io is about the size of Earth's Moon, and when astronomers looked to our Moon, they saw a cratered, geologically silent world. Most analysts expected to find craters on the orange Jovian moon. What they got instead was volcanoes, lots of them.

Why would Io be the most volcanically active moon in the Solar System, while our Moon, a twin in size, is essentially dead? The answer lies in a force called tidal friction. The gravity of nearby Jupiter pushes and pulls on Io, as does the gravity from the large moons Europa and Ganymede.

These three Galilean moons (Io, Europa and Ganymede) are in resonance, with Ganymede circling Jupiter a quarter of the way around for every Io orbit (a 4:1 resonance), and Europa making the trip half as many times for each Ionian circuit (2:1). This cosmic taffy pull causes the surface of the rocky moon to rise and fall by some 80 m each day, heating the interior and resulting in the most powerful known volcanoes in our Solar System.

Io's volcanoes completely resurface its globe roughly every three centuries, painting its tawny landscape in spectacular color. Silicates, sulfur, sulfur dioxide frosts and other volatiles blanket the little moon's volcanic landscape. Research has shown that sulfur dioxide transitions through a series of brilliant color changes as it cools. Molten material is most often black, changing as it cools into red, orange and yellow. Sulfur dioxide frost also powders the landscape in blue and white. Sulfurous frosts rim many vents and fissures in snowy hues. Some areas are even green, earning the moniker "golf courses." Mountain peaks and fault cliffs punctuate Io's colorful plains here and there, but for the most part, the moon is remarkably flat. Even the majority of its volcanoes build up fairly gentle slopes around themselves, or erupt directly from fissures and vents in the floor of the surrounding plains.

The surface area of Io is equivalent to the continents of North and South America combined. But the tiny moon's surface is encrusted with more volcanoes than we see on the entire surface of Earth, including its undersea volcanoes. One sulfurous lake, called Loki, is a volcanic crater – or caldera – filled with a black lava lake some 200 km across. Within Loki's tarry liquid is a gigantic iceberg-like island covered by frozen sulfur dioxide. The outcrop at lake's center is about the size of Rhode Island.

The longest lava flow, called Amirani, races across 300 km of Ionian real estate. It is as long as the Mississippi River delta. Many flows end in cracks, from which geysers tower hundreds of kilometers into the airless void. The geysers on Io are incredibly vigorous, in part because of Io's low gravity. Volcanologist Susan Kiefer estimates that if Yellowstone's Old Faithful geyser

Fig. 3.3 Features on Jupiter's remarkable moon Io include the lava lake Loki (left, compared to the U. S. state of Rhode Island, inset), the 290-km-tall eruptive plume from Tvashtar next to the white silhouette of Belgium, and the great doughnut of dark material surrounding the volcano Pele, here seen next to the country of France (Images left to right, courtesy of NASA/JPL; NASA/JPL/ JHUAPL; NASA/JPL/Caltech)

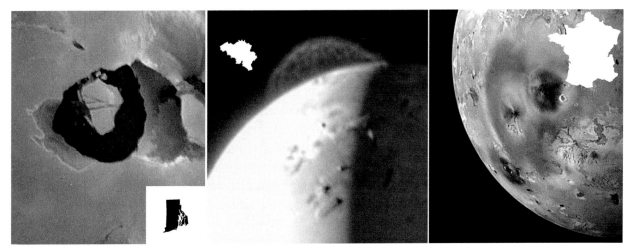

was transported to the low gravity and near vacuum of Io, the 50-m-high gush seen by Earthly tourists would blossom to a 35-km altitude.

The highest, most dramatic plumes arise from a type of volcanic activity known as Pillanian eruptions. Pillanian volcanoes spew the highest plumes seen anywhere in the Solar System. The most powerful eruption ever recorded came from the volcano called Surt. The event was witnessed remotely from the Keck telescope on February 22, 2001. During the eruption, Surt's output bested the average power of all other Ionian volcanoes. Its incandescent eruption covered some 1900 km, an area larger than the Los Angeles metropolitan region.

Each second, 100,000 tons of sodium and sulfur dioxide belch from Io's vents. Much of this volcanic debris escapes into space, forming a great donut-shaped cloud around Jupiter, tracing its course along Io's orbit. Jupiter's strong magnetic fields charge the sulfur into a frenzy of ions, electrically linking the little moon to the giant planet. Two trillion watts of electricity crackle through the sodium cloud, power equal to 350 nuclear power plants. So much for a small, quiet moon.

EUROPA

The moon next door to Io, Europa circles Jupiter further out. It is also small – about the size of Earth's Moon – and scientists expected it to be cratered. Instead, its remarkable surface is glazed in water-ice, etched by long dark lines unlike any other world in the Solar System. Like rocky Io,[1] almost no craters have been spotted on the brilliant blue-white orb. But unlike Io, Europa's surface has no stony mountains or rock-strewn valleys. Its surface appears to be pristine frozen water with areas of darkening, perhaps stained from the sulfur drifting in from Io, or from organics rising from beneath. Under its ice veneer is a 100-km-deep ocean of salt water. For comparison, the deepest spot in Earth's oceans is the Challenger Deep in the Marianas Trench, diving to some 10,994[2] m below the surface of the ocean (roughly a tenth the depth of Europa's estimated abyss).

Europa is truly an ocean moon, containing more water within its tiny sphere than all the water on Earth. Europa's water would create a ball of liquid 1754 km across, some 483 km in diameter wider than one made by Earth's lakes, seas and oceans. Despite its chilling Jovian temperatures (−220 °C at Europa's poles), Europa retains its water in liquid form, for the same reason that Io has its volcanoes: tidal friction. At Europa, the force is less, but enough to warm its interior.

Europa's subsurface ocean gives itself away in a subtle magnetic field around the moon. Like Earth's oceans of salt water, Europa generates a weak magnetic field, a telltale sign that deep waters slosh within. This hidden brine sculpts the landscape above it in varied and distinct forms. Ruler-straight parallel stripes streak across the frozen landscape, bracketed by long parallel ridges rising hundreds of meters into the ebony sky.

[1] In fact, no impact craters have been located on Io. Says Io volcano expert Rosaly Lopes, "We don't see any partially degraded craters or crater rims. The surface is very young, as continuous volcanic activity has modified it, covering all the signs of old craters."

[2] With an accuracy of ±40 m.

The ubiquitous cracks, called linea, provide Europa with its cracked-eggshell appearance. The bands run over Europa's face for thousands of kilometers, draping the surface in a specific directional pattern. In the northern hemisphere, they trend in a northwestern direction, while southern fractures trend southwesterly. Some of the features crack along great arcs. The directional tendencies of Europa's remarkable "highway system" suggest that its global ice crust is not linked to its core, but rather is free-floating. The moon's ridged plains have fractured into vast sections of ice rafts. Areas referred to as chaos regions seem to have collapsed into a sea-like slurry. Within this frozen mash-up, ice rafts and bergs have shifted and rotated before freezing solid again. These terrains have frozen into place after separating like gigantic puzzle pieces that remain close enough for scientists to understand their original pattern.

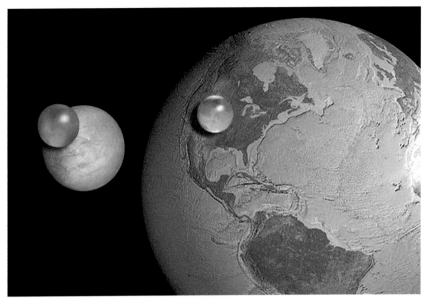

Fig. 3.4 *Europa's entire inventory of water would equal a sphere 483 km wider than that from all of Earth's oceans and lakes combined (Image © Jack Cook, Woods Hole Oceanographic Institute and Kevin Hand, JPL. Used with permission)*

At the bottom of Europa's global ocean lies a rocky seafloor where active volcanic activity – in the form of hydrothermal vents – may seed the water with minerals. For scientists searching for life on other worlds, this is a significant phenomenon. Volcanic sites on Earth's seafloor host colonies of unique biomes. Cut off from the sunlight that sustains Earth's complex web of life, the cold depths were once thought to be a desolate, high-pressure desert swathed in eternal darkness. But minerals carried by hydrothermal vent eruptions provide nourishment for an entire deep-sea menagerie. Bacteria in this gloomy world are completely independent of any food sources related to solar energy. Instead, they rely on sulfur for nutrition. Some of the microbes associated with vents known as "black smokers" are thought to be similar to the most ancient life on Earth. These archaeobacteria provide the foundation for an alien food chain that includes such exotic creatures as frizz-covered Pompeii worms, blind crabs, giant tubeworms, one-eyed shrimp and other strange creatures living in these frigid pressure-filled environments.

The discovery of deep-sea colonies at hydrothermal vent sites has invigorated the astrobiology community. Researchers who study the possibility of life on other worlds have not missed the parallels between the terrestrial ocean floor and the conditions that may exist beneath the alien oceans of Europa. This gives astrobiologists hope that, like the seafloor volcanoes of Earth, life may thrive in the depths of Europa's primordial seas.

Fig. 3.5 Two high-resolution images of Europa's rugged Conamara Chaos, a region of collapsed and refrozen ice crust. In the top *image, the cliffs are 100 m high, roughly the height of Florence's El Duomo cathedral. At* bottom, *flat plates give way to jumbled ice. The mountain rises 250 m above the frozen tumult. Brooklyn Bridge just spans the fracture along the left side (Europa Galileo images courtesy of NASA/ JPL. El Duomo is the author's photo. Brooklyn Bridge modified from private collection, used with permission)*

GANYMEDE AND CALLISTO

Next out from Jupiter are two behemoths as moons go. Ganymede, brackered by Callisto and Europa, is actually 383 km wider than the planet Mercury, at 5262 km in diameter. Callisto is the most remote of the four Galileans, with a diameter of 4820 km. Both Ganymede and Callisto are cratered worlds, but Ganymede falls under the influence of those same tidal forces made famous by Io and Europa, while distant Callisto has remained geologically silent since the earliest years of our Solar System's history.

As size affects the very nature of a cosmic object, Ganymede constitutes a hybrid of sorts. Smaller than Earth but larger than any other moon, Ganymede is a planet in its own right. It is the only moon known to generate an internal magnetic field. This type of phenomenon is due to Ganymede's core, which may contain molten iron, much like we will see in the rocky planets (Mercury, Venus, Earth, the Moon and Mars). Spacecraft have also found evidence of an induced magnetic field like Europa's, suggesting a deep internal ocean of saltwater. The densities of both Ganymede and Callisto suggest that the two large moons are made up of about 60 % rock and 40 % water. How much of that water is ice, and how much –if any – constitutes a hidden sea, is open for debate. Because of the massive amounts of ice in their makeup, the overall density of the moons is less than Earth's own, creating gravity fields of about 1/7 that of Earth's gravity.

Like that of its siblings Europa and Io, Ganymede's geology is made complex by the influence of Jupiter and the nearby moons. Its face displays a marked contrast in regions: dark, ancient terrain resembling the landscapes of Callisto, while the bright regions are somewhat like Europa. Ganymede seems frozen in its evolution, trapped in limbo halfway between Callisto and Europa.

Ganymede's bright Europa-like provinces transition from its dark regions abruptly, often at the edges of cliffs. Line upon line of icy ridges and cliffs rise out of the bright ice. Craters pepper its surface, but far fewer than in the dark terrain. Parallel valleys and ridges, called grooved

Fig. 3.6 The closest view of Ganymede's peculiar terrain obtained by the Galileo orbiter. Three Olympic pools end-to-end lend familiar scale to the alien scene (Image courtesy of NASA/JPL)

terrain, plow across Ganymede's bright regions. Tens of kilometers wide, these folded tracks of ice run across the face of Ganymede for hundreds of kilometers, slicing swaths through the dark terrain, breaking it into gigantic polygons. Ganymede's bright grooves appear to be scars of planetary expansion. Like a baking biscuit, Ganymede swelled as it froze, cracking its surface. The composition of Ganymede's bright terrain is almost pure water-ice, suggesting cryovolcanism or flooding from the interior, but its ridges and valleys indicate that tectonic forces are the culprit.

Like all planets and moons, Ganymede has been battered by asteroids, comets and meteors. Impacts have scarred Ganymede's face on many scales, leaving features ranging from small bowls a few feet across to the vast impact basin of Gilgamesh, some 600 km in diameter, covering an area slightly larger than the country of Turkey. Small craters sometimes dig out dark floors or cast rays of dark material. Across many of Ganymede's frozen plains, crater ghosts, known as palimpsests,[3] leave a shadowy bright imprint beneath newer craters and ridges.

CALLISTO

Callisto's battered face is geologically the oldest of the Galileans, with ancient craters probably dating back to the Late Heavy Bombardment, some 3.8 billion years ago. This face betrays an inert interior. Callisto has no hint of Io's volcanoes, no suggestion of Europa's ridged fractures, none of Ganymede's bright etchings. Callisto is its own world.

Unlike the other Galileans, Callisto's guts remain as jumbled as they were at its inception. Differentiation, the process of heavy material sinking to create a dense metallic/rocky core, took place on all the other Galileans. But gravity studies show that Callisto is a scrambled blend of ice and rock.

Callisto's tattered surface bears witness to a landscape virtually unchanged from within, sculpted by a drizzle of meteoroids, comets and asteroids since its surface first solidified. But its landscape has its own unique landmarks: great pinnacles and knobs of water-ice rear up from its chocolate-brown plains. The bizarre, several hundred meter-high ice pinnacles most often form from degraded crater rims. In places, the landscape appears to have disintegrated, chewing away at crater rims and cliff walls.

[3] Palimpsest is a word that pertains to ancient manuscripts. When authors wrote on parchment – a rarity – they would often reuse the parchment, scraping off the old ink with a mixture of vinegar and milk. But a ghost image of the older text would often remain, a valuable source of information for those who study ancient manuscripts.

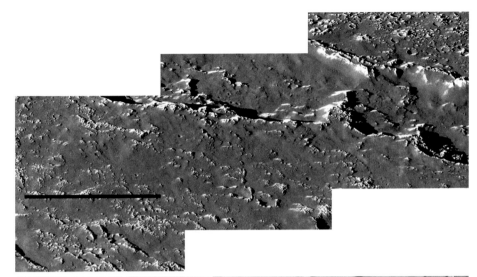

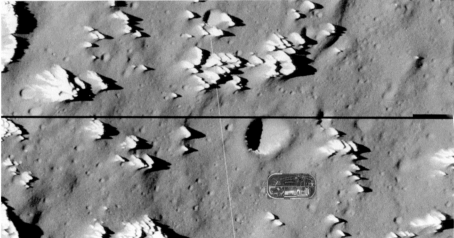

Fig. 3.7 The ancient craters of Callisto have degraded into rings of icy promontories and pillars, as seen in the detail below. The approximately 1000-m-long Indianapolis Motor Speedway provides scale (Callisto image courtesy of NASA/JPL. Speedway via Google Earth)

WEIRD MOONS BEYOND JUPITER

The size of the Galilean moons, along with their proximity to Jupiter and to each other, determines their character. But beyond Jupiter, where temperatures plummet even further, we find more outlandish satellites. These moons seem to ignore their own size, displaying exotic geology despite diminutive scales.

First on our tour is the unusual irregular moon of Saturn called **Hyperion**. Hyperion is easily large enough to be round, measuring 360 km at its widest. But the bizarre little ice chunk is quite irregular. Its midsection is 266 km wide, and its poles are separated by just 205 km, the distance from London to Birmingham. In addition to its odd shape, the moon is not tidally locked. Any irregular moon should point its longest axis toward its parent planet, but Hyperion cartwheels dramatically as it orbits Saturn. Its tumbling movement is essentially chaotic, something unseen on other bodies of its size (although similar chaotic movement takes place among the small satellites of Pluto).[4] These irregularities suggest that the moon has been chipped away by violent impacts. Its cratered surface takes on the appearance of a giant sponge, and its low density indicates that the spongy appearance is more than skin deep; the ice moon is remarkably porous. Hyperion's craters are unlike craters on other moons. Its impact scars seem to have compacted, crushing the surface and expanding the craters into deep pits. An unknown substance fills the interiors of many pits, and landslides have raced down the slopes of some crater walls.

One of the smallest of the distant weird worlds inhabits an orbit around the planet Uranus. **Miranda** is a scant 470 km across, just 1/7 that of Earth's Moon. Nevertheless, its surface has been folded and mixed by titanic forces. Why all this activity on such a small globe? Theories range from a nearly annihilating asteroid impact to unusual internal processes involving ammonia cryovolcanism.

Less than half the satellite has been seen in detail,[5] during the barnstorming flyby of *Voyager 2* in 1986. The hemisphere seen is blighted in three areas by crumpled, oblong zones called coronae. These terrains are unlike anything seen elsewhere in our Solar System. Spreading up to 300 km across, the coronae incorporate a series of concentric ridges and troughs. Research suggests that the coronae formed over plumes of rising material from Miranda's core. As material spread out and invaded the surface, it may have triggered local episodes of cryovolcanism. There are features indicative of flood deposits oozing from fractures at several sites. In other areas, ice has flooded the ground, obliterating ancient craters. Within many of the coronae regions, ridges intersect in chevron arrangements. The most dramatic of these lies within the center of Inverness Corona. Here, a bright checkmark of ridges spanning 100 km engraves the smoother ridges surrounding it. Faults cut across it. One of the faults, Verona Rupes, reaches far into the cratered terrain. With near-vertical slopes stretching along 19 km of steep real estate (a depth ten times that of the Grand Canyon of the Colorado), Verona Rupes stands as one of the most spectacular crags in the Solar System.

Just next door to Miranda orbits **Ariel**. Next out in distance, Ariel is essentially a twin in size to the nearby moon Umbriel. Each is roughly 1150 km in diameter. But while their sizes are comparable, their natures differ dramatically from each other. Umbriel's dark surface appears to be uniformly cratered, with a primitive landscape that has changed little from the early days of the moon's development. This is exactly what we would expect in a small moon. But surprisingly, Ariel's old cratered plains crown mesas that rise above bizarre canyon floors. Floods of sinuous material have resurfaced the floors of these gorges. These frozen streams bow up in the center, and winding troughs run down them, typical of some lava flows on Earth. Surface flows appear to be water "lavas" that have erupted across the canyon floors, spilling out onto adjacent plains and craters.

[4] Hyperion is in a 3:4 resonance with Titan, which may also contribute to its wild dance around Saturn.

[5] During the *Voyager 2* flyby of the Uranian system, the Sun illuminated only the southern hemispheres of the planet and its moons.

Something has resurfaced the countryside outside of Ariel's canyons, too, perhaps very recently in the moon's geological story. The moon's ice is the brightest – and newest – of all the major Uranian satellites. Ariel's face appears to have been almost completely resurfaced sometime after the initial bombardment era of our Solar System, whose rain of asteroids and comets tailed out about 3.9 billion years ago. Craters are sparse in many regions, and its surface lacks many 100-km craters, confirming a geologic age younger than its battered siblings Umbriel and Oberon. Although Ariel checks in at just 1160 km, it underlines the fact that small is not necessarily quiet. In the outer Solar System, the link between size and geologic activity is skewed, as seen later in an even smaller moon of Saturn.

Enceladus shines as the brightest moon in our Solar System. This snowy orb circles the planet Saturn at a distance of 238,000 km, making the trek once each 33 h. Its diameter is only 34 km greater than Miranda's, but its nature couldn't be more dissimilar. This tiny ball of ice is big news. It's one of the most volcanically active worlds in the Solar System, perhaps second only to Io and Earth. But the geysers of Enceladus are fundamentally different. They are nearly pure water, jetting some 500 km into the skies above the frosty landscape. They erupt from a series of parallel canyons called Tiger Stripes. The Saturn-orbiting Cassini spacecraft has charted nitrogen, methane, ammonia and carbon dioxide within the gossamer plumes. It also caught a whiff of elaborate carbon-rich molecules. Something complicated is going on in the chemistry beneath that ice. What's more, the ice particles in the plumes contain sodium chloride (ordinary table salt) and other salts. Salty ice is difficult to make unless it is flash-frozen from saltwater. The geysers are rooted in an ocean centered at Enceladus's south pole.

Fig. 3.8 The crags of Miranda's Verona Rupes would tower above those of the Grand Canyon of Colorado. Verona Rupes's rim stands ten times as high as the Grand Canyon's tallest cliffs (Grand Canyon panorama by Danny Santiago via Wikipedia Commons: https://commons. wikimedia.org/wiki/ File:Cedar_Ridge,_Grand_ Canyon.jpg. Art by the author.)

Fig. 3.9 The Uranian moons Ariel (behind) and Miranda float beside the Arabian Peninsula (Photos courtesy of NASA/JPL)

Like Ariel and Miranda, craters blanket some areas of Enceladus, while other regions have been completely smoothed over by floods of material. Still other areas are cracked and twisted by canyons. A fine, reflective powder dusts everything, a snow from the continual particle blizzard of the eruptions. Researchers estimate that geysers on Enceladus dump 150 kg of water into space each second.

We now come to two satellites that do a fine job of imitating the Death Star in *Star Wars,* one large and one smaller. **Mimas** is a diminutive moon just 396 km wide (slightly smaller than Enceladus), and is the smallest moon in hydrostatic equilibrium (spherical from its own gravity), probably due to the fact that it consists mainly of water, with a very small rocky component at center. Despite its small size, it sports a deep crater, called Herschel, nearly a third the diameter of the moon itself. Herschel's raised rim rises 5 km up, while the crater floor dips to 10 km below the surrounding plains. Experts believe the impact came close to shattering the small worldlet. At the center of the crater, a central peak rears up 6 km from the crater's floor. If Mimas were the size of Earth, Herschel would cover an area greater than Australia.

Perhaps more than any other moon, Mimas has a profound effect on the rings of Saturn. Mimas's gravitational influence empties the material from the Cassini Division, the great open pathway between Saturn's A and B Rings. Mimas doesn't empty the Cassini Division completely, but its particles are sparser than those in the adjacent rings. Despite appearances, the region is filled with fine dust. Within that dust, a clearing opens. It's called the Huygens Gap. The particles in the Huygens Gap orbit at a 2:1

resonance with Mimas; for each circuit that the little moon travels around the giant planet, these particles make two turns. This resonance forces them to migrate outward, leaving the gap. Another clearing between the C and B Rings is also in resonance with Mimas, this one at a ratio of 3:1.

As if that was not enough, the convivial Mimas is in relationship with other moons. It shares a 2:1 resonance with its much larger sibling Tethys, and a 2:3 resonance with the outer F Ring shepherd moon Pandora.

Saturn's moon **Tethys** shares something else in common with Mimas – it also sports a great wound from an ancient impact. The magnificent crater Odysseus is so large that the country of Nepal would fit comfortably inside it. Its interior flooded by water as it formed, creating a crater floor that curves with the rest of the moon's surface. With a span of 445 km, the great bowl extends two-fifths the diameter of the 1060-km moon. Nearly directly opposite of Odysseus, on the far side of the moon, spreads a vast smooth, lightly cratered plain. Shockwaves from the titanic impact may well have flattened the opposite hemisphere. Nearby, and working its way ¾ the distance around the globe, a great chasma called Ithaca rends the surface. The gorge is 100 km across, 3 km deep and over 2000 km long, making it nearly ten times the length of the Grand Canyon of the Colorado. If draped across Europe, Ithaca Chasma would stretch from Lisbon to Copenhagen. Like Mimas, the modern face of Tethys is a battered, geologically dead landscape.

Climbing up our bizarre moon scale, we find **Iapetus**, a two-toned Saturnian moon measuring 1460 km in diameter. Iapetus is the most distant of the regular mid-sized moons of Saturn. Although it is the planet's third-largest moon, Iapetus orbits much farther from Saturn than the other major moons. Like all large moons but Hyperion, Iapetus is tidally locked, always keeping the same face toward Saturn. The satellite's orbit is inclined by 15 ½° to Saturn's equator. An observer standing on its rugged surface would see Saturn in one place in the sky, never rising or setting. But because of Iapetus's tilted orbit, Saturn would appear to wobble like a top, slowly pirouetting once over the course of its 2-week "day." At Iapetus's range, Saturn would spread as far across the sky as three full Moons in Earth's sky.

Iapetus' trailing hemisphere is as bright as dirty snow, while its leading face is nearly as dark as asphalt. Infalling material from Saturn's dark outer moon Phoebe may be peppering the dark hemisphere. But the boundary between hemispheres is far too sharply defined to have been caused by simple dark "snowfall." Something else is afoot. In the transition zone between dark and light, scattered pools and curtains of dark materials fill in valleys and craters. It is a very thin layer, in many places less than a meter deep. Some small impacts have pierced through to bright ice beneath. Scientists have come to the conclusion that a complex set of processes contributes to the piebald nature of the moon. Water-ice migrates from illuminated, warmer areas, such as sun-facing crater walls, to nearby shadowed areas that are colder. The ice leaves behind a lag of darker material, while brightening the shadowed terrain.

Fig. 3.10 *The gravitational pull of several moons creates gaps in Saturn's rings. Mimas, the socialite of moons, has several gravitational relationships (Note that although their locations are correct, the moons are not shown to scale) (Art by the author)*

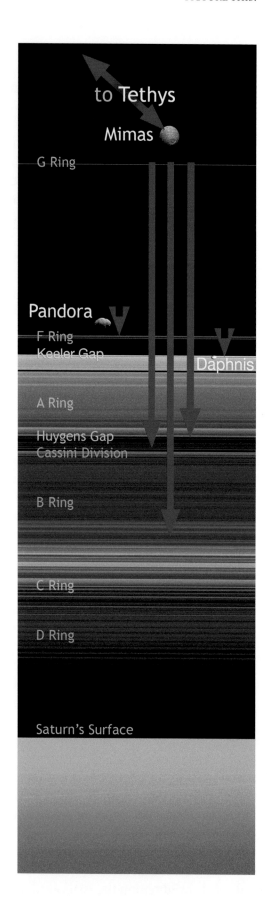

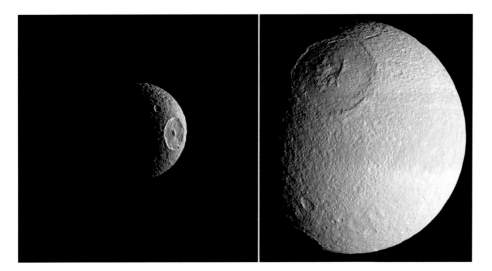

Fig. 3.11 A single giant crater blemishes the face of the small moon Mimas (left) and another indents the larger Tethys. The two moons are shown in their correct relative sizes. In both cases, the impacts probably came close to destroying the moons (Image courtesy of NASA/JPL/SSI)

Aside from its remarkable split personality, Iapetus has another wonder, an enormous ridge that stretches along the equator like the seam on a cheap rubber ball. The ridge towers some 13 km high – half again the height of Mt. Everest – and spreads as wide as 20 km. The spectacular ridge branches into isolated peaks on one end. The peaks making up portions of the ridge are among the tallest mountains found on any planet or moon. The ridge is heavily cratered, implying that it formed early in Iapetus' history.

The Iapetus Ridge is one of the most perplexing features in the entire Solar System, as nothing remotely like it is found on any other planet or moon. Theories abound. Some researchers suggest that the ridge was forced up due to an abrupt slowing of Iapetus' rotation. But why did the ridge form only on the equator, unlike the tectonic features found all over Europa and Ganymede due to similar forces?

One of the first theories proposed that as it formed, Iapetus was spinning so rapidly that centrifugal force pushed it up at the equator, but this theory does not seem to fit with Iapetus's internal structure. Others suggest that the ridge was thrust up from forces below, but this would require Iapetus's stiff outer layer to be relatively thin. In fact, the moon's crust is thick enough that it holds up those tall mountains without deforming around them as it would were it thinner. This leads some analysts to conclude that it did not emerge from the interior, but rather was somehow deposited onto a solid, thick crust. Several researchers have recently proposed that a debris ring around Iapetus – the left-overs of a small moon – gradually collapsed upon the equatorial zone.

Any small moon captured by a larger body settles into an elliptical orbit, but over time tidal forces would circularize its path. A small moon circling Iapetus may have spiraled ever closer to the yin/yang moon. Eventually, the gravity of Iapetus[6] pulled the moon apart, scattering it into a thick ring of boulders, gravel, sand and dust. The ring flattened and

[6] Iapetus' gravity is just 1/40 that of Earth's, but still enough to destroy a moon that gets too close.

became aligned to the equator (just as Saturn's is aligned to its own). Because of Iapetus's low gravity, particles rained down at glancing blows and very low speed. The theory proposes that the shower of stony debris built up material rather than excavating craters and destroying the surface. The recently detected debris ring around the asteroid 10,199 Chariklo (see Chap. 2) may provide us with a window into the past of Saturn's mysterious, dappled moon.

TITAN

In our progression of weird Saturn moons, we have yet to visit the strangest moon yet, Titan. Like a planet out of place, Titan looms large in the outer Solar System. Many researchers consider it to be a target just as important as Mars for future exploration.

Titan measures nearly the size of the planet Mercury. Its 5150-km diameter makes the moon second only in size to Jupiter's Ganymede. Because of its size, Titan is able to hold onto a deep atmosphere, maintaining an air pressure 1.6 times as dense as Earth's at sea level. Its atmosphere is composed primarily of nitrogen, as is Earth's.

Fig. 3.12 The planet-sized moons Titan (above) and Ganymede (left) compared to the planet Mars. Of the three, Titan has the densest atmosphere (Photos courtesy of NASA/JPL. Art by the author)

Titan's weather is among the most alien of the worlds with atmospheres. Winds move in great atmospheric tidal waves across the face of the moon, dragging sand dunes into parallel lines hundreds of kilometers long. Titan has the closest thing to Earth's hydrological cycle that we have seen on any solid-surfaced world, but where Earth's rain is comprised of water, Titan substitutes cryogenic methane. Like water on Earth, methane cycles regularly, raining out of the sky, carving rivers and pooling as lakes and seas, then evaporating as vapor to return to the sky once again. Water on Titan makes up the planet's ground, frozen hard as stone in Titan's −179 °C temperatures.

NASA and the European Space Agency combined forces to launch the massive Cassini/Huygens spacecraft in the fall of 1997. Settling into orbit around Saturn in July of 2004, the Cassini/Huygens mission has revolutionized our knowledge of the gloomy moon. Cassini's instruments are sensitive to the near-infrared light that seeps through Titan's fog. Cassini is also equipped with radar, which can resolve fine surface detail. With nearly each loop around Saturn, Cassini coasts by Titan, turning its radar toward the surface to map a thin strip tens of kilometers wide and hundreds of kilometers long.

In addition to ethane and butane, backyard grills would be well supplied with a shower of "natural gas" fresh from the orange skies above. The rain precipitates out of methane-nitrogen clouds 20 km up. Methane humidity increases closer to the poles, where methane/ethane lakes and seas have been found. Titan's weather may resemble the most severe Earthly drought conditions for periods of centuries or millennia, broken by brief periods of flash flooding. Titan's rainstorms may resemble Earthly monsoons, falling sporadically, but with ferocity.

Titan displays desiccated equatorial regions covered in vast seas of sand dunes and dry mountains. The Belet Sand Sea alone is over 3000 km across, covering an area of 3.3 million sq. kg. The extent of this equatorial dune field rivals the Arabian Desert.[7]

Some research indicates that the dunes – one of the few Earthlike formations seen on the bizarre moon – may consist of pulverized water-ice. But even more alien processes may be at work. A constant rain of organic material drizzles from Titan's ruddy sky. This sooty sludge may pile up into material that is blown into the dune-like formations visible in Cassini's images.

Europe's Huygens Titan probe piggybacked aboard the Cassini orbiter. The coffee table-sized craft was a spectacular success, radioing data over a period of 2 h and 27 min as it descended through Titan's extensive atmosphere. Surprisingly, the probe continued to transmit from the surface to Cassini for another 69 min. Huygens actually survived on the surface for a total of 3 h and 14 min, long after the mother craft was no longer close enough to relay good data. The tale Huygens told of Titan's weather was intriguing. Much of what we know about Titan's methane rains and atmospheric structure came from this plucky European spacecraft. Huygens was able to sample gases directly from its surroundings as it

[7] The Arabian Desert covers an area 2.33 million km sq.

Fig. 3.13 A thin strip of Titan's Belet Sand Sea next to the Arabian desert at the same scale. Belet extends farther to the north and south than seen in this radar swath (Titan image courtesy of NASA/JPL/SSI. Arabian Desert image courtesy of NASA)

descended on a small parachute. It charted moisture, both in the air and within the surface material. *In situ* measurements confirmed the existence of a complex organic chemistry, showing that the giant moon's chemical brew results in bizarre prebiotic conditions that may have parallels with ancient Earth. Titan's conditions are not all about its size, but its mass contributes to the great atmosphere it holds.

TRITON

While Saturn's Enceladus belches streams of water vapor, another kind of geyser erupts on the distant moon Triton, largest satellite of Neptune. Triton is about the size of Earth's Moon, but like the other weird moons we've visited, it appears to have a surface nearly devoid of craters.[8] Its face has been completely resurfaced in the recent past, and likely continues to be transformed.

Triton is an interloper. It came from somewhere else, most probably somewhere out in the Kuiper Belt. Evidence includes the fact that the large moon orbits Neptune in a retrograde direction, opposite Neptune's direction of spin. If Triton had formed as part of the Neptunian system, its orbit should take it in the other direction, revolving *with* the planet and any of

[8] In the most detailed Voyager images, 179 features have been identified as possible impact craters. Of those, many are indistinct or possibly of volcanic origin.

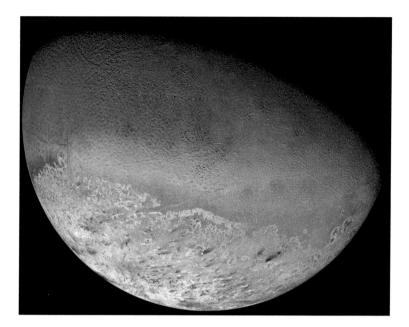

Fig. 3.14 *The bizarrely beautiful moon Triton, in orbit around Neptune, is frosted in pink nitrogen ice (Image courtesy of NASA/JPL)*

its native moons. Another odd aspect of Triton's orbit is that it is inclined to Neptune's equator by a steep 157°, unlike the other moons, which formed within the system. Additionally, in comparison to the moon families at Jupiter, Saturn and Uranus, Triton inhabits an orbit in the same relative region around Neptune that should be inhabited by a family of major satellites. Looking at the other three giant planets, researchers see patterns of moons in varying orbits. But there is only one moon in what should be a busy region around Neptune – Mimas-sized Proteus. Another major moon, much farther out, is the 350-km rock Nereid. This outermost moon of Neptune has the most eccentric orbit of any known moon, ranging from 5.5 to 1.4 million km. The muddled orbits of Neptune's family of satellites may be the aftershock of a violent early encounter in which Triton passed near enough to Neptune to be captured. This fierce interaction would have left Triton orbiting in its oddball direction, and would have destroyed or ejected many of the major moons from Neptune's system, leaving the remaining satellites in peculiar orbits. With time, Triton's orbit settled down into the more circular pathway it follows today, leaving behind a scarcity of other major moons.

Active eruptions send dark plumes of soot into Triton's extremely thin atmosphere. Two of the plumes spotted by the *Voyager 2* spacecraft measured 8 km high. At their tops, the plumes hit a jet stream, where they stretch into 150 km-long trails of material drifting down into long curtains, staining the surface. Other images of Triton's southern polar region revealed more than 100 dark, streaky deposits, pointing preferentially northeast, away from the southern polar ice cap. The streaks stretch from tens to hundreds of kilometers across the surface, implying that plume activity must be fairly common.

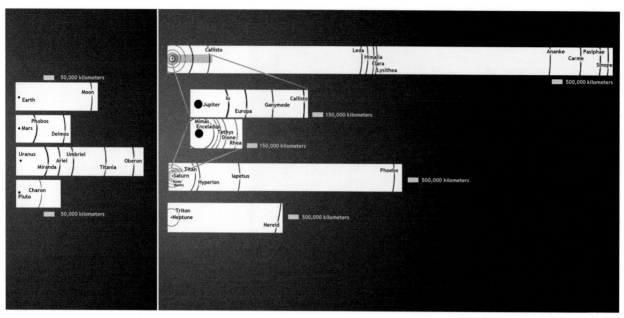

Fig. 3.15 *The scale of moon systems, compared. Moons circle their parent planets at distances that stretch – at least for the outer planets – to millions of kilometers. To show all the selected moons (not all moons are shown) three scales are used (note the scale bars). At right, we see detail insets of the inner moon systems of Jupiter and Saturn (Art by the author)*

The cryovolcanism on Triton differs significantly from cryovolcanism on the satellites of Jupiter and Saturn. The plumes appear to be similar to geysers, but their mechanism is a mystery. One model proposes that sunlight passes through the clear, pink-tinted nitrogen ice, and is heated as it is trapped within the ice, in a sort of solid version of the greenhouse effect. Heat builds up inside, sublimating the nitrogen directly from ice into nitrogen gas. The vapor pressure builds up and finally explodes into the near-vacuum of Triton's air. Whatever their cause, the geysers of Triton rank among the most spectacular scenes in planetary science.

With an overview of the weirdest moons, from tiny rocks to massive planet-like worlds, we can see the influences of size and distance. But to really understand how these scales play into the character of the various members in our solar family, we must venture to the planets themselves.

Chapter 4
A Tour of the Planets

Captain Kirk stands at his console. He's in trouble again, and needs to consult with Star Fleet Command back on Earth. Despite the fact that the *Enterprise* is 20,000 light-years from Earth, Kirk is able to talk, in real time, to his superiors, which ultimately enables him to save the universe. Kirk is able to ignore the great distances in the galaxy by the use of subspace radio, a clever invention engineered by… Hollywood. But the truth is that we have no such technology, and when we communicate with our spacecraft emissaries throughout the Solar System, their distances are underlined by the long time lag in our transmissions to and from them. When a spacecraft – for example, *Voyager 2* – is out at Neptune, still well inside our cosmic backyard by Captain Kirk's standards, one-way communication takes just under 4 h.[1] That is, the time it takes light (or radio waves) to travel from Earth to Neptune at its closest point in its orbit is roughly 3 h and 54 min. It is a very long way to Neptune.

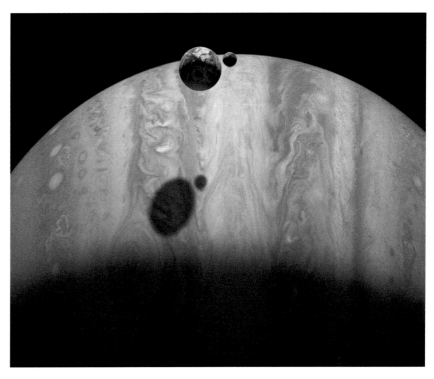

Fig. 4.1 Jupiter dwarfs our Earth and Moon (Photos courtesy of NASA.)

DISTANCE AND TRAVEL

How long? As we saw in our introduction, the planets are arranged in such a way that the gaps between them increase as their distances from the Sun grow. Earth, circling the Sun at a distance of 93 million miles, is one astronomical unit (AU) from the Sun. The terrestrial (Earth-like) planets huddle around our Sun in orbits ranging from Mercury's .39 AU to Mars' 1.5 AU (meaning that Mars orbits 1½ times as far from the Sun as Earth does). Beyond the Red Planet, in the realm of the giant worlds, the distances fan out into vast gaps. Jupiter loops the Sun over five times as far as Earth, at 5.2 AU. Saturn nearly doubles that distance at 9.5 AU. Uranus and Neptune make their circuits in the outer darkness at nearly 20 AU and 30 AU, respectively. Communication across those vast distances is measured not in seconds but in minutes or hours. At its closest approach, light and radio time to Mars is over 12 min, and a message to Saturn – one way – would take over an hour (see chart). Engineers at the Jet Propulsion

[1] When Voyager encountered Neptune, the two planets happened to be near each other in their orbital trek, both on the same side of the Sun. Had Neptune been opposite the Earth, Voyager's 246 min light-time would have stretched to some 8 h.

Light times to the planets

Planet	Distance from sun (in astronomical units)	Light travel time
Mercury	0.39	193 s or 3.2 min
Venus	0.72	360 s or 6.0 min
Earth	1.00	499 s or 8.3 min
Mars	1.52	760 s or 12.6 min
Jupiter	5.20	2595 s or 43.2 min
Saturn	9.54	4759 s or 1.3 h
Uranus	19.82	9575 s or 2.6 h
Neptune	30.09	14,998 s or 4.1 h
Pluto (KBO)	39.44	19,680 s or 5.5 h

Laboratory do not have conversations with the Mars rover Curiosity. They have communications sessions, blocks of data that they send, and blocks of data that come back from the rover after a delay of many minutes.[2] Without their magical subspace radio, our Star Trek heroes would certainly be frustrated, subjecting their viewers to endless commercials between action scenes.[3]

Our textbooks and magazines display the planets as nicely arranged marbles on a disk of concentric circles. This is a simplified way of looking at our planetary neighborhood, made necessary by the size differences of the worlds and their distances from each other. For example, if we arrange the tiny terrestrial planets on a page in such a way as to render Earth at 3 cm from the Sun, Neptune would require a foldout nearly a meter long. With Earth's orbit as a circle 3 cm across, mighty Jupiter would travel a circular path 28 cm across.

Traversing such great distances is daunting. Only a handful of spacecraft have ventured across the Asteroid Belt to reconnoiter the gloomy outer Solar System, and these missions have presented engineering challenges of the highest order. A direct flight to Mars, in the best-case scenario, requires a travel time of roughly 6 months. A direct flight to Saturn could take nearly a decade. But the movement of the spheres and their gravitational influence work in concert to gift spacecraft planners with a shortcut[4] across the great void. It's called "gravity assist."

Gravity assist was incorporated into the flight of the Jupiter-bound Galileo spacecraft, a massive orbiter carrying an atmospheric probe. Galileo took a long, circuitous route to the king of worlds, coasting by Venus, picking up speed for a long loop out beyond Earth. But the craft was not yet traveling fast enough to make it out to Jupiter. It needed two more gravity boosts, and these both came from Earth flybys. Along the way, flight engineers also targeted the craft to fly by two asteroids, making good use of its prolonged Solar System tour. The *Pioneer 11* spacecraft took a leisurely six and a half years to reach Saturn. Its speed was increased by the gravity of Jupiter, which acted as a slingshot to toss the little robot further into space.

Voyager 1 took only 3 years and 2 months to cover the same distance, again using the gravity of Jupiter to fling it Saturnward. Its sister craft, *Voyager 2*, spent 4 years en route, because its flight path was crafted to carry it farther, on to Uranus. Since it had a small window in which to pass, engineers sacrificed speed so that Saturn's gravity could precisely bend Voyager's trajectory toward Uranus. Voyager used the same technique at Uranus, sailing by the green giant at just the right angle to continue on a path to Neptune. Its voyage to Neptune was shortened by many years with the use of these gravity assist maneuvers. Neither Voyager would have

[2] Or even hours, depending on the timing of orbiting relay stations such as the Mars Reconnaissance Orbiter or ESA's Mars Express.

[3] Even sending a radio message to our own Moon next door takes effort. At a distance of 240,000 miles away, radio signals take 1½ s to travel the one-way distance.

[4] In terms of travel time, but not necessarily distance.

made it beyond Jupiter on its own power. Launched aboard a powerful Titan III/Centaur, the twin explorers would have settled into quite different fates without a gravity assist from Jupiter. Timing was everything. Had Jupiter not been in the right spot at the time of flyby, both robots would have ended up in orbits reaching a high point from the Sun of about 5 AU. Their circling paths would have then taken them diving down to a spot closer to the Sun than Earth's orbit. They would have been trapped in that long loop, circling the Sun for eons, until a passing comet or planet disturbed their flight.

More recently, the Cassini spacecraft cashed in on two Venus flybys, one Earth encounter and one pass by Jupiter to get it out to the ringed giant. Cassini is a massive spacecraft the size of a school bus. No booster was large enough to send it directly, so it had to rely on the gravitational power of the planets.

SIZING UP THE WORLDS

While distance diagrams can be visually misleading, the diameters of the planets themselves can also be deceiving. Saturn's radius is just slightly less than Jupiter's (by roughly 19,000 km), but the Ringed Planet is far less dense. In fact, Jupiter is by far the densest of all worlds. If the entire family of planets and moons were placed on one side of a scale, with Jupiter on the other, Jupiter would still weigh more.

All of these worlds spin on their axes while revolving around the Sun, each world at its own speed. At its equator Earth turns at more than 1600 km/h. Thanks to its gravity, we don't feel it, but we do get the sense of movement if we are aware of the Sun's path across the sky each day. Earth itself travels at more than 107,826 km/h in its great circle around the Sun. The Sun, in turn, is moving around the center of our Milky Way Galaxy at a rate of 724,205 km each hour. But with all this movement, we perceive our world as a stable place with air above and solid ground below. Even the fact that Earth is a big ball of rock didn't dawn on most ancient people. A select few noticed that ships disappear over a curved horizon. Pythagoras of Samos pointed out that southbound sailors saw the southern constellations rise progressively higher in the night sky, while the Sun appeared to migrate northward. This could only happen, he reasoned, if Earth was a sphere. As it turns out, he was right, and he figured this out in the sixth century B.C.

But how big is this ball? Actually living on it makes it a little challenging to tell. About 250 years after Pythagoras made his claims, a Greek philosopher/mathematician named Eratosthenes attempted to measure the world using two not very long sticks. He didn't gauge Earth's size with the sticks, but rather watched their shadows. At local noon in what is today the city of Aswan, the Sun cast no shadow from directly overhead. But at local noon in Eratostheses' own home of Alexandria, further north, a stick did cast a shadow. By comparing the differences, the mathematician was able to estimate Earth's size as 14,500 km, just 1750 km off.[5]

Our home world is a pretty big deal to us. We are surrounded by vast oceans as far as the eye can see, towering mountains that challenge the greatest athletes, immense deserts that spread across Earth's equator and poles. And yet, our entire world could be swallowed twice over by Jupiter's famous cyclone, the Great Red Spot.

Let's visit a few of these worlds, and some of their landmarks, to get some idea of our Solar System's cosmic scale. We'll start close by. Although our Moon is the size of Pluto or several of the outer planet moons, it's close to home, rocky like the inner worlds, and familiar, gazing down at us from the "heavens." Though not a planet, it's a good place to get our bearings.

EARTH'S MOON

The heavens used to be so simple. The first-century writer Paul spoke of three "heavens": the first was where the birds and clouds floated; the second was where the stars wheeled; and the third was where God lived. To Ptolemy and other ancient observers, the physical universe could be broken down a bit further. Earth, whether flat or round (a source of debate in ancient times) sat in a stationary spot at the center of all things, while the birds, sunsets, clouds and rainbows existed in a sort of middle limbo between Earth and a star-embossed crystalline sphere, perfect in order and structure, as were the wandering stars (planets), Sun and Moon. Galileo's little telescope changed all that. The Moon, it seemed, was no perfect orb, but rather had rugged mountains and circular hollows. These hollows are craters, and craters make a fine yardstick for geologists who are determining the relative ages of planetary surfaces.

We saw in Chap. 1 that even Earth has craters (for a little historical background on our view of craters, see Chap. 7 in this book). A hail of comets, asteroids and meteors has battered every planet and moon in existence, from the torrential rains of metal and stone during the formation of the Solar System to a gentler meteorfall today (gentler in rate, but not necessarily in potential violence). A moon or planet that is geologically dormant or inactive will be covered in craters to the extent that when a meteor hits, the craters it forms obliterate the same number of craters. This surface is referred to as "crater saturated." It is so covered in impact scars that the total number of craters cannot be increased. Some of the small moons in the outer Solar System, such as Mimas, are crater saturated. Other planets have so much weather or geological upheaval that their surfaces are nearly or completely devoid of craters.

Earth's moon falls somewhere in between. Fresh impact craters pock its face here and there, while lava flows or avalanches cover older blemishes. Some impacts have draped spectacular rays of bright debris across hundreds of miles. One such crater, one of the brightest spots on the Moon's visible face, is the crater Tycho.

[5] Earlier, Plato had estimated Earth's diameter at 2250 km.

Tycho is 85 km in diameter. It is among the youngest of the giant bright rayed craters on the Moon's near side, with well-preserved walls and a central peak. Tycho's bright rays, ejected debris from the titanic impact, drape across over 2000 km in every direction from the crater. Within the crater walls, cross sections of the Moon's stratigraphy are preserved as layer upon layer of differing material. Ponds of impact melt are scattered across the crater floor. Samples of impact glass returned from the *Apollo 17* landing site, which lies within one of these rays, have been radiometrically dated at 108 million years old. If these samples are from the Tycho impact event, they suggest that Tycho was created 108 million years ago, but precise dating will have to await samples from the crater itself.

In almost any huge impact crater, when the impact from a meteor vaporizes the rocky or icy surface, the molten surface rebounds, forming a central peak in the center. Tycho's spectacular central peak rises 2 km above the crater floor, and spreads out across 15 km at its base. Atop the peak rests a boulder even larger than the one seen in Fig. 4.2. "You can see, in the bedrock that is exposed on the central peak, that there is layering in the rock that was uplifted," explains David Kring of the Lunar and Planetary Institute. "The rock has different shades…You get some sense that the crust of the Moon is complex, and if you were to actually go there, you could collect the samples and get a sense of what those colors mean."

The peak consists of material that has rebounded back up after being compressed in the impact, like a drop of water splashing up after a pebble dropped into a pond. Although it is a lofty mountain today, its material actually came from a greater depth than the floor of Tycho. The crater floor is plastered by impact melt – rocks that turned to liquid from the heat of the impact event, and then surged across the floor of the crater in a tsunami of molten stone. Impact melt flowed downhill along crater walls,

Fig. 4.2 Left: *The bright crater Tycho has the most elegant and extensive rays of any crater in the southern lunar highlands (Image courtesy of Gregory Revera, Wikimedia commons: https://commons.wikimedia.org/wiki/File:FullMoon2010.jpg).* Center: *A portrait of the crater taken by the Lunar Reconnaissance Orbiter, with a field of view some 110 km across. Note the* arrow *pointing to the rock seen in the next image. This gigantic boulder was probably tossed up during the violent process of impact that created Tycho (Image courtesy of NASA/GFSC/Arizona State University)* Right: *Close-up of the enormous boulder compared to a standard U. S. football field (Image courtesy of NASA/GFSC/Arizona State University, overlay by the author.)*

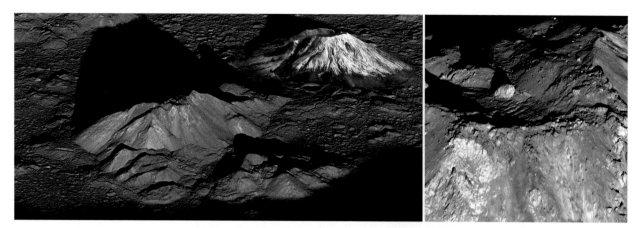

Fig. 4.3 Left: *The central peak of Tycho rises nearly 2 km high, an altitude slightly higher than the prominence of Mt. St. Helens, plopped down behind it (Mt. St. Helens photo courtesy of Garry Hayes, used with permission). Right: On the mountaintop, a titanic boulder may be a fragment of the meteor that created the crater, or a block of the ancient lunar crust. Our football field is again spread across the area for reference (Lunar Reconnaissance Orbiter images courtesy of NASA/GFSC/Arizona State University.)*

pooling at the base of the cliffs as it cooled. Kring says, "That image tells you that all that information is just waiting. There is bedrock and you can go pick up a piece. You can see the fault scarps, you can see all their features. You can see scarps where the melt has actually flowed down the central peak. You can see cracks in the impact melt pool that are centered on blocks of rock that were caught in the flow, [causing] cracks around them like spokes in a wheel."

The distances to the other planets, where the secrets of the Solar System's history lie in the ground and on its surfaces, underline the uniqueness and proximity of our own Moon. Kring points out that, "The Moon is its own world with its own history, and it's only three bloody days away. You can get to the Moon faster than you can get to some parts of Earth." And being so close, it's a natural target for going in search of those pieces of the Solar System chronicle. The entire first 500 million years of Earth history has been largely erased by weather, volcanoes and the shifting of Earth's continents. But that history is sitting on the lunar surface, displayed like a library. And like all planetary science, the Moon enables us to understand our own world better. "The Moon has this window into history that is missing elsewhere."

As planets and moons go, our own natural satellite is quite large compared to Earth. In fact we have the largest moon compared to a primary planet. (Pluto's Charon also subtends a significant percentage of its parent planet's diameter, but these tiny worlds fall into the ice dwarf class; see Chap. 2 in this book.). The worldlet next door offers a stabilizing effect to Earth's axis. As a planet spins, it tends to wobble – or *precess* – just as a spinning top wobbles on a table. Mars, whose axial tilt (or obliquity) is nearly equal to Earth's 23.44 °, may slowly tip over to spin nearly on its side 50,000 years from now. But our heavy Moon, circling Earth's equator, dampens this wobbling effect. This is critical to life on Earth, as the planet's tilt determines climate and seasons.

Our unique Moon also causes the tides, which link to many of life's cycles. This global slosh rejuvenates the waters, while the ocean currents help to drive weather systems. Not only do the oceans rise and fall with the lunar cycle, but the solid ground itself does. Earth's plate tectonics, which are unique in the Solar System, may have been triggered or spurred on by the Moon's gravitational influence.[6] Earth's plates, on which the continents rest, move along a stony conveyor belt on the surface of Earth at about the rate that the human fingernail grows. As our jigsaw-puzzle plates of rock collide into each other, mountain ranges arise. (The Himalayas are the product of the Indian plate moving north and crashing into the Asian plate.) In other areas, plates subduct, or dive under each other. As the sinking rock melts, it releases chemically imprisoned gases back into the atmosphere, often through volcanic eruptions. Without plate tectonics, Earth's minerals would end up in the ocean basins, the carbon cycle would cease, and the planet's atmosphere would chemically starve. The volcanoes of Mars died out long ago, in part due to a lack of this tectonic system. Again, this geological difference may be due, in part, to scale. Smaller Mars had less heat to begin with, and heat[7] is what drives Earth's tectonic plates to move. With less inner warmth, the crust of Mars probably became immobile early in its history.

The Moon also was probably responsible for pulling away a great deal of the greenhouse gases surrounding the primordial Earth, leaving the planet with a more benign environment than the planet next door, Venus. Venus is a twin in size to Earth, but its dense poisonous atmosphere may be left over from its formative years, when the planet had no large Moon to help clear the gases away. Such are the benefits of a large moon nearby. In fact, Earth's Moon is just 1400 km smaller in diameter than the smallest planet, Mercury.

MERCURY

The closest world to the Sun is also the runt of the planetary family. Mercury is a world of contrasts. Its daytime temperatures soar to 450 °C, but ice wallows in crater floors at the poles. Because Mercury has no atmosphere to distribute its heat, shadowed areas may drop to −170 °C, a temperature swing of 600 °C!

Although small, Mercury is dense, with an iron core larger, in comparison to its diameter, than any other planet. Research indicates that a planet's magnetic fields, if it has them, are generated by a combination of a fairly rapid spin and molten iron surrounding the core. The planet's spin sets up currents within the molten core, generating magnetism. Mercury spins at a leisurely rate of once each 59 days. Nevertheless, the planet somehow produces its own magnetic fields. Just one percent as strong as Earth's, Mercury's fields are still able to grab the nearby Sun's stream of charged particles, swirling them into hurricanes of plasma that slam into the barren surface.

[6] Mars may have had something similar to plate tectonics at one time. Magnetic patterns in Martian rock seem to contain repeating patterns similar to those on Earth's ocean floors where the seafloor has spread apart. If so, this process ended long ago. Jupiter's moon Europa also appears to have had similar movements across regions of its ice crust, but these motions are related more closely to sea ice drift.

[7] Liquid water may also play an important role.

Mercury races around the Sun in just 88 days. Because its rotation is 59 days long, the Mercurian skies see a strange series of sunsets and sunrises. From some locations, the Sun appears to rise for just a few days, then set where it rose, then rise again to make its long trek across the sky. At sunset, the reverse scene plays out as the Sun sets, rises briefly, and sets again. If you miss your first romantic sunset, just wait a few days.

As Mercury cooled during its formation, it appears to have shrunk, losing as much as 14 km at its equator. This compression caused the surface to wrinkle, leaving mile high lobe-shaped scarps that wander for hundreds of kilometers across the cratered landscape. Researchers refer to these cliffs as rupes.

About 4 billion years ago, a scant 500 million years after Mercury condensed from the great cloud surrounding the Sun, an asteroid 60 miles across plowed into the infant planet. The power of its impact equaled 1 trillion megatons, as powerful as 20 trillion Hiroshima bombs. The cosmic intruder created a vast impact basin roughly 960 miles (1550 km) wide. Known as the Caloris Basin, this elegant, multi-ringed arena could hold the entire state of Texas.

Not all of the small world's craters came from asteroids and meteors. The planet also shows evidence of volcanic craters – or calderas – with a number of flows pouring from them. These flowing plains cover many of the ancient craters. Close-ups from the Mercury-orbiting MESSENGER spacecraft reveal that some of the vents have punched through craters with ages of 3.5–1 billion years, which means volcanoes continued to vent throughout that time period. A bizarre landform called the hollows is peculiar to Mercury. These shallow, irregular depressions usually show

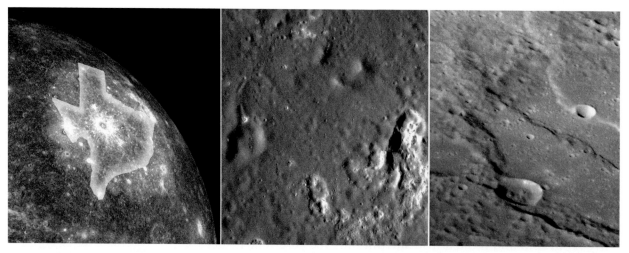

Fig. 4.4 Left: *The great Caloris impact basin with the state of Texas superimposed.* Center: *"Hollows" are small volcanic depressions encircled by bright halos, seen here in the floor of Mistral Crater. Mercury also plays host to larger volcanic vents, seen here as irregular sunken features with blankets of ash issuing from them.* Right: *One of Mercury's wrinkle ridges, or rupes. These scarps run across the Tharkur region, which was shoved together from top right and bottom left as the planet cooled and shrank. The image is just under 60 km across, the distance from Mannheim to Darmstadt, Germany (All Mercury images from MESSENGER spacecraft courtesy NASA/ Johns Hopkins University Applied Physics Laboratory/Carnegie Institution of Washington.)*

themselves in the floors of craters or valleys. They seem to be associated with explosive venting of volcanic gases. Like other geological forces, volcanoes can be dependent on the size of a planet, as we will see at the Hadean world next door, Venus.

VENUS

That important rule of thumb in planetary science – the idea that the larger the planet is, the more likely it is to have geologic forces at work within it – is alive and well on Venus. While rules are meant to be broken (as we saw with Europa, Enceladus and Io), the trend holds true for the second planet out from the Sun.

In size, Venus is Earth's twin, but in nature it is its antithesis. Our home world's weather mixes our atmosphere well, and our daily spin of 24 h gives us subdued temperatures between day and night. On Venus, temperatures soar to nearly 900 °F, while it turns lazily once each 243 Earth days. As a year on Venus lasts just 225 days, a Venusian day is longer than a Venusian year. This means that the planet is actually turning in a retrograde direction, opposite to most things in the Solar System. And although Venusian gravity is just a bit less than that of Earth (Venus is less dense), walking through the air would be like walking at the bottom of a swimming pool. The atmosphere is 90 times as dense as Earth's at sea level.

Where Earth is awash in liquid water oceans, Venus is desiccated. Volcanic flows, fractures and summits grace the face of this Dantean world. Fully 90 % of the surface features on Venus are related to volcanism, and they are big. The longest canyon in the Solar System, Baltis Valles, spans 2 km wide at places and over 6000 km long. At one time, this meandering valley flowed with molten rock. Newer lava streams have obliterated both ends of the dried-up lava river; it may have been considerably longer. As it stands today, Baltis is as long as China's Yellow River.[8]

Venusian volcanoes also rank near the top in size. Some Venusian mountains have grown to Everest size, towering 7925 m into the sulfuric acid haze. They pepper the plains and rest atop the highlands. But perhaps the strangest of the Venusian erupters are the pancake domes. These eerie formations are roughly disk-shaped and flat to slightly domed on top. They average 24 km in diameter and 760 m in height. Their flat tops are often fractured. Over 150 have been identified on the planet. These vast disks remind us that even on a twin to Earth, the scale of familiar forms may be dramatically different.

The hellish conditions on Venus afford us another lesson in the distance scale. Venus made famous the term "greenhouse effect," which contributes to the planet's balmy 485 °C surface temperatures. "Why's Venus so hot?" asks astrophysicist Jeff Bennett. "People will tell you it's because it's closer to the Sun, and it is. But when you do it to scale, you can see that it doesn't matter." Bennett refers to a scale model of the Solar System on

[8] The Yellow River is the 6th longest river in the world, but Baltis Valles could easily have rivaled our longest, the Nile (6650 km).

Fig. 4.5 This Magellan radar mosaic, centered at 12.3° north latitude and 8.3° east longitude, extends across 250 km of ground in the Eistla region of Venus. The disk-like pancake volcanic domes are 65 km across. Here, they are compared to the lights of the greater Los Angeles/San Diego area of southern California (Photos courtesy of NASA.)

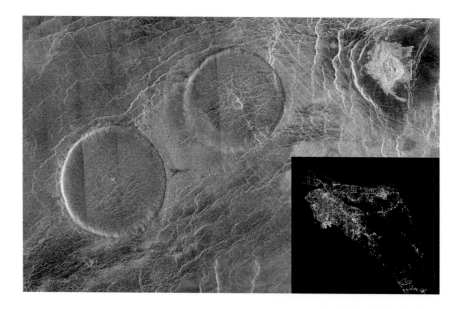

the campus of the University of Colorado in Boulder, a model that begins with the Sun at CU's Fiske Planetarium. The Sun is the size of a grapefruit, mounted to a granite pyramid, with Venus 11 m away and Earth at about 15 m. "If you put a bonfire at the Sun on that pyramid in front of Fiske Planetarium, now go stand where Earth is, and then go stand where Venus is. Yes, it's going to be warmer when you stand at Venus, but it's not going to make the difference between 15 and 470 °C. You can see there's got to be something else going on there. Distance isn't enough to explain it. If you haven't done scale, it seems reasonable to say the explanation is that Venus is closer to the Sun, but that's only a small part of the story." Scale provides that understanding.

MARS

Mars is a small world with big features. Although its diameter is half that of our home world, volcanoes twice as tall as the tallest mountains on Earth rise from its plains, and its equator is split by a grand canyon as long as the continental United States. When astronauts finally arrive there, they will be exploring territory as vast as all Earth's continents put together.

Thanks to advanced orbiting observatories such as ESA's Mars Express, India's Mars Orbiter Mission, and NASA's Mars Reconnaissance Orbiter (MRO), and to our surface probes like NASA's Opportunity and Curiosity rover, we have scrutinized the Red Planet up close. From orbit, we see boulders less than a meter across; with this resolution we have been able to locate our landers and rovers, including ones thought lost to history. Recent orbital images revealed the fate of the European Space Agency's *Beagle 2* lander. Like so many Mars explorers throughout history, the lander disappeared during landing, on Christmas Day of 2003. Searches by

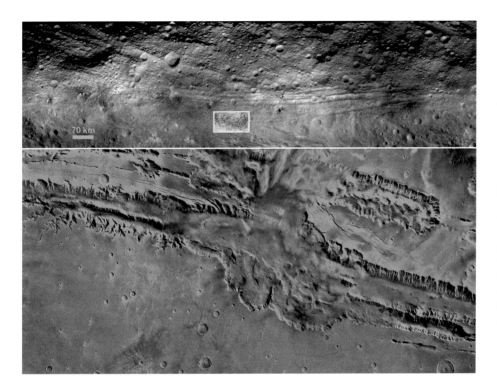

Fig. 4.6 An equatorial canyon system some 370 km long spreads across the surface of the asteroid Vesta, above. The inset shows a map of the American Grand Canyon to the same scale. (Note that the serpentine Grand Canyon itself bends and meanders, running a total of 443 km.) Below is Valles Marineris, grand canyon of Mars, also to the same scale (Grand Canyon map ca. 1926 via Wikipedia Commons; https:// commons.wikimedia.org/ wiki/File:Map_of_the_Grand_ Canyon_National_ Park_1926.jpg planetary images courtesy of NASA/ JPL. Art by the author.)

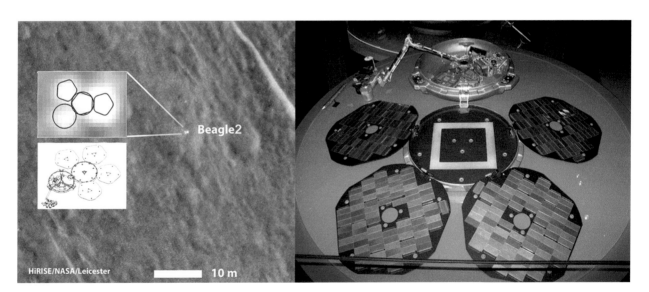

Fig. 4.7 The British Mars lander Beagle 2 as seen by NASA/JPL's Mars Reconnaissance Orbiter. The failed lander appears to have only partially deployed, with three of its five petals showing. The ideal configuration is seen in the mockup at right (Images courtesy of NASA/JPL-Caltech/University of Arizona/University of Leicester; Beagle mockup via Wikipedia Commons; https://commons.wikimedia. org/wiki/File:Beagle_2_replica.jpg)

several orbiters proved fruitless, but after years of searching, NASA's Mars Reconnaissance Orbiter resolved the little lander, partially deployed, in the equatorial sands of Isidis Planitia. The lander is only 2 m across, so it was at the limits of resolution.

The incredibly detailed images by modern spacecraft reveal Mars to be a geologically active, exciting world. Dry river valleys meander from rugged lowlands in the south to vast low-lying plains in the north, where past rivers may have emptied into large bodies of water. Along the margin of the plains, formations resembling rocky benches appear to mark the limits of lakes or seas. In many locations, minerals associated with standing water pepper the landscape, a testament to a warmer, wetter Mars in ancient times. Large debris aprons – perhaps filled with ice – fan out along Martian valleys, while newly-formed craters excavate fresh ice from below the surface. Orbiters have charted active gullies moistened by liquid saltwater. Other crater images reveal complex histories of changing climate, wind and water erosion.

One such crater recently hosted a terrestrial visitor. NASA's Opportunity rover studied the Victoria Crater in Mars' Meridiani Planum region from September 2006 to August 2008. Victoria is named after Victoria City in the Seychelles. Mission planners informally named its various coves and outcrops after the bays and peninsulas explored by Ferdinand Magellan. The naming of features has more than entertainment value; it aids the navigation team in communicating about complex routes that the rover will be commanded to execute.

In Fig. 4.8, we see Victoria Crater. This view, taken by MRO from orbit, includes other craters, as well as boulders and dunes on the crater floor, but these visual clues to scale are insufficient to judge the crater's size. For scale, we compare it to the Roman Coliseum. At the crater rim, seen by NASA's Opportunity rover (below), even these structures are difficult to judge until we put an intrepid geologist next to the cliff face.

As we see with Victoria, the scale of smaller craters on various planets and moons is difficult to discern just by looking. Some bowl-shaped "simple" craters a few meters across appear identical to ones kilometers across (see Fig. 4.9).

While small craters share similar shapes, craters begin to take on new characteristics once they venture beyond a certain diameter. Larger craters cannot retain a bowl shape, as their molten rock rebounds to form a central peak (as did the crater Tycho). These complex impact features may sometimes develop terraced rims. Still larger impact sites form concentric rings as they expand from the classification of crater to impact basin. The floors of larger craters also begin to flatten out into interior plains. And while details vary, these characteristics can be seen on rocky worlds as well as icy ones.

The structure of some Martian features echoes counterparts on Earth, but not always in scale. In Fig. 4.10, we see a phenomenon called patterned ground, or ice polygons. The ground in tundra areas on Earth, where ice is present just below the surface, melts and heaves into these unique polygonal forms. On Mars, in the polar regions, this polygonal fracturing is also present, indicating the freezing and thawing water just beneath. But on Mars, the polygons are ten times the size. This may be a result of the long seasons on Mars, twice the length of those on Earth. On a human scale, the Martian landscape can be hauntingly familiar. The Phoenix lander set down in the Martian arctic and returned photos remarkably similar to the

Fig. 4.8 Victoria Crater from above, compared to Rome's Coliseum and seen from the surface. The geologist is there only for scale. Actually putting him there is far too expensive (Coliseum photo courtesy of Richard Dennis Kortum, used with permission. Art by the author.)

Fig. 4.9 Each crater in this sequence is slightly more than ten times the size of the one to its left, beginning with a 390-m simple crater and ending with the ~600-km impact basin Mare Moscoviense on the Moon's far side (Images courtesy of NASA.)

Fig. 4.10 Two views of the Martian arctic (above) compared to similar views on Earth. At upper left, the Phoenix lander snapped this shot of its surroundings on the Martian north pole. Below it is a similar view of the Canadian arctic. At upper right, the patterned ground seen from Martian orbit, and below it, a photo taken out an airplane window above Kotzebue, Alaska (Note the difference in the scale bars.) (All photos courtesy of NASA except Alaska, which is by the author.)

Canadian tundra. Despite rusty daylight skies, blue sunsets, two moons and frigid temperatures, scenes like these remind us that our cosmic neighbor next door is the most Earthlike world in the Solar System.

THE GAS AND ICE GIANTS

When we look out from Earth across our planetary system as a whole, we discover that our terrestrial neighborhood is a small part of the big picture. Carolyn Porco, head of the Saturn Cassini imaging team, explains, "No matter how you measure it, whether you count the number of bodies, whether you add up the amount of mass, or whether you calculate the volume taken up by the orbits of those bodies, the vast majority of our Solar System lies out beyond the orbit of the asteroids. Inside it is just a bit of flotsam." It's hard to argue the dominance of the outer system. While the terrestrial planets huddle within 1.3 AU from the Sun, Neptune stretches to a distance 30 times that. Early on, astronomers grouped Jupiter and Saturn with Uranus and Neptune, referring to the quartet as "gas giant planets." And they are giants, combing a volume equivalent to 2205 Earths.[9] But modern research has revealed fundamental differences between the four. When a planet gets to the size of these behemoths, strange things

[9] 1321 Earths would fit inside Jupiter, 764 for Saturn, 63 for Uranus and 57 for Neptune.

begin to happen at their cores. The core pressures of Jupiter and Saturn – the gas giants – are so great that hydrogen becomes a liquid metal.[10] The cores of ice giants Uranus and Neptune, much smaller, consist of rock and ice at lower pressure and temperature.

The gas and ice giant planets are worlds of weather. Unlike their terrestrial siblings, their surfaces are not ice and rock, but rather gases, winds and clouds. Their cores are far denser than Earth's core, but as one moves out from the midpoint of a giant planet, the environment transitions from a rocky and metallic solid to a liquid, then to a gas. There is no clear demarcation between solid and gas, no place to stand.

The gravity of the giant worlds is so strong that they were able to hold on to even the lightest primordial gases that originally gathered to form them (referred to as reducing atmospheres). The ratios of helium and hydrogen within both Jupiter and Saturn are similar to the makeup of the Sun itself. While hydrogen and helium rule the atmospheres of all four giants, ammonia makes up a large constituent of the atmospheres of

[10] In fact, pressures become so great that raining carbon particles may transform into house-sized diamonds deep within these colossal worlds.

Fig. 4.11 The gas giants Jupiter and Saturn sit behind the ice giants, Uranus and Neptune. Near the front, Earth and the Moon seem miniscule in front of the outer planets (Planet images modified from NASA/JPL sources. Art by the author.)

Jupiter and Saturn. Uranus and Neptune have evolved slightly different mixes of gases, with more methane and less ammonia. Methane enriches not only their air, but also their color. It is methane that causes the blue-green hues of Uranus and Neptune.

The giant planetary quartet displays common patterns and fundamental differences. All four worlds have bands of clouds encircling them, parallel to the equator. The cloud bands are wracked by incredibly strong winds, but they remain stable within their respective latitudes. On Earth, continents block airflow, causing pressure waves that mix things up. Storms come and go on the order of days or, in the case of hurricanes and monsoons, weeks. But the cloud banners and giant storms of the outer planets may last for decades or even centuries. Astronomers have observed Jupiter's Great Red Spot for nearly 400 years.

The cloud banding on the outer planets is most obvious on Jupiter and Saturn. Despite its distance from the Sun, Neptune's bands are nearly as well defined. We see subtler striping on Uranus, whose axis is tilted so that the planet essentially rolls around the Sun on its side. While Uranus's belts are more subdued than those of its siblings, they are distinctive, with some broken into patterns similar to giant chain links (Fig. 4.12).

The king of worlds, Jupiter is by far the largest object in our Solar System with the exception of our Sun. More than a thousand Earths could fit within its sphere. Oval storms the size of Earth's continents spin like hungry maws, eating away at the clouds around them. Banners of feathery vapors stream for hundreds of miles, pulled along by hurricane-force jet streams.

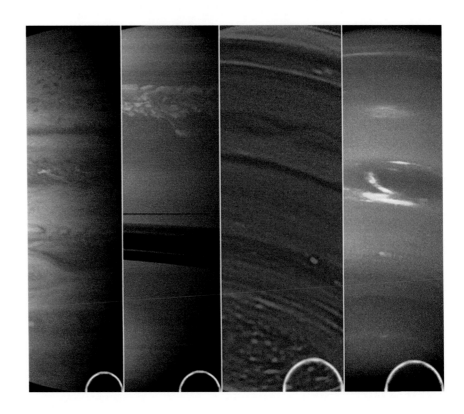

Fig. 4.12 Cloud formations of the four giant worlds compared, all scaled so that the planets are identical in size. Left to right: Jupiter, Saturn, Uranus and Neptune. Circles indicate the size of Earth compared to each world (Photo credits left to right: NASA/JPL/SSI; NASA/JPL/SSI; Keck Telescope, University of Wisconsin, Larry Srmovsky; NASA/JPL.)

Ponds of rich color mingle in a continual interplay of color and shifting texture. Farther below, a great plain of reddish ammonium hydrosulfide cloud spreads out in a jagged fen to a distant horizon. Here and there, pale blue-gray clouds break through the ruddy cloudscape, water clouds boiling up from beneath. The water cloud deck is the lowest, floating nearly 100 miles below the highest ammonia cirrus. The vertical distance between these two cloud decks is deeper than the entire atmosphere of Earth. Underneath this cloud layer, rain falls into an eternal night of crushing pressures and searing temperatures, as the hydrogen making up most of the air around us transforms into a liquid, and then into a fluid metal.

Saturn, too, has belts and zones of clouds, but it glows in a golden patina, a color due primarily to a layer of high altitude haze. If we were to fly within the clouds of Saturn, the horizon would be so vast that it would appear perfectly flat. Towering white clouds billow up into the deep blue-purple sky. Canyons snake through the clouds below, reddish-brown flows tinged by ruddy organic molecules. Within those canyons, titanic lightning bolts powerful enough to supply a small city for a week, flash and crackle through the airy chasms. And beyond it all, etching parallel lines across the sky, the glorious rings of Saturn glow against the sky, accentuated here and there by the miniature crescents of their tiny shepherd moons. Saturn is blasted by supersonic winds, among the fiercest in the Solar System.

The poles of Saturn display wondrous features. In the south, a great vortex pulls in concentric cloud bands like a titanic whirlpool. The storm's walled rim rises 40–65 km high. Despite its 560 kph winds, the storm remains locked directly over the south pole. Across the northern hemisphere, a colossal hexagon blankets territory as far across as two Earths. The geometric stream of air is stable and long-lived. The hexagon's cause is not well understood, and adds to the uniqueness of the golden giant Saturn.

The lower cloud decks on Uranus may resemble blue Neptune in color, but they are buried deep in the Uranian atmosphere, beneath a layer of brownish hydrocarbon smog. Many narrow cloud bands encircle the planet near the equator. *Voyager* spotted only a few discrete clouds, most blazing along on the equatorial winds at up to 580 kph. The visible clouds consist of methane ice crystals, and may be welling up from below.

In Neptune's clear upper atmosphere, temperatures hover at about −220 °C. Descending from this level, temperatures increase with depth. At a pressure comparable to Earth at sea level, temperatures rise to −200 °C.

Despite the fleeting nature of its clouds, Neptune does have a few long-lived features. One of the first features *Voyager 2* resolved in far encounter shots was a large blue storm reminiscent of Jupiter's Great Red Spot. (But unlike Jupiter's cyclone, Neptune's spot disappeared within a few years.) The storm was approximately the same size relative to the planet, a deep pool floating in an azure sea of clouds. Scientists christened it the "Great Dark Spot," a reference to Jupiter's Great Red Spot. The storm spanned the distance of Earth's diameter.

Fig. 4.13 The mysterious hexagon at the north pole of Saturn dwarfs our home planet (Saturn image courtesy of NASA/JPL/SSI. Earth image courtesy of Goddard Spaceflight Center/NASA. Art by the author.)

Although considerably bested by the gas giants, Uranus and Neptune could still swallow about 60 Earths each. While the blue-green behemoths are essentially twins in size, they are quite different in nature. When *Voyager 2* carried out the first close reconnaissance of the planet in 1986, one member of the press quipped that Uranus held the distinction of being the "most boring planet." Compared to Jupiter, Saturn and Neptune, weather on Uranus is subdued. This may be the result of its internal heat. The gas giants and Neptune put out more heat than they receive from the Sun. Not so Uranus. Its temperature appears to be in balance with the incoming solar energy, resulting in an atmosphere that is less mixed and more sedate.

In Neptune's clear upper atmosphere, temperatures hover at about −220 °C. Descending from this level, temperatures increase with depth. At a pressure comparable to Earth at sea level, temperatures rise to −200 °C.

Despite the fleeting nature of its clouds, Neptune does have a few long-lived features. One of the first features *Voyager 2* resolved in far encounter shots was a large blue storm reminiscent of Jupiter's Great Red Spot. (But unlike Jupiter's cyclone, Neptune's spot disappeared within a few years.) The storm was approximately the same size relative to the planet, a deep pool floating in an azure sea of clouds. Scientists christened it the "Great Dark Spot," a reference to Jupiter's Great Red Spot. The storm spanned the distance of Earth's diameter.

The lower cloud decks on Uranus may resemble blue Neptune in color, but they are buried deep in the Uranian atmosphere, beneath a layer of brownish hydrocarbon smog. Many narrow cloud bands encircle the planet near the equator. *Voyager* spotted only a few discrete clouds, most blazing along on the equatorial winds at up to 360 mph. The visible clouds consist of methane ice crystals, and may be welling up from below.

Voyager left us with a portrait of a cold, distant world simmering in serenity. And so the portrait remained, until astronomers began a 1993 campaign using the Hubble Space Telescope to search for moons at Uranus and Neptune. The overexposed shots revealed moons, but some of the underexposed images uncovered something else – complex cloud systems. Uranus was now past solstice, so more of the planet was illuminated. Twenty years after *Voyager*, clouds began appearing in the northern hemisphere, the face that was finally getting illuminated after half a century of night. By 2007, astronomers were tracking more clouds than *Voyager* saw during its entire encounter. Bands became well-defined, and dark storms lapped each other around the planet, merging into larger storms. Hubble images revealed a dark spot similar to Jupiter's Great Red Spot, spanning an area two-thirds the length of the continental United States.

The Uranian weather patterns bore similarity to those of the sapphire world Neptune. Its clouds are intrinsically bluer than those of Uranus. They contain some kind of coloring agent that actually tints them. Dark belts and glowing storm clouds paint the face of the blue leviathan, while around it circle unique ring arcs and a host of moons (see Chap. 3).

Neptune's internal heat makes for the dynamic activity in its skies. Tendrils of brilliant methane ice crystals stretch out for hundreds of miles, while smaller versions skate over the deep blue lower cloud deck in hours or days. Blooms of whitish clouds skitter across subtle belts and zones, sometimes spiraling into cyclonic features.

Neptune's weather systems are far less ordered than those of Jupiter and Saturn. The zones seem to drift and cross from one latitude to another. The gas giants display cloudy belts that remain consistent, as if painted onto the surfaces of the planets. Changes occur on yearly scales with only occasional abrupt shifts. But on Neptune, the changes are dynamic and constant. The zones are ill defined, and the wispy clouds seem to come and go haphazardly

In Neptune's clear upper atmosphere, temperatures hover at about −360 °F. Descending from this level, temperatures increase with depth. At a pressure comparable to Earth at sea level, temperatures rise to −334 °F. (Again a switch from one measuring system to another.)

Despite the fleeting nature of its clouds, Neptune does have a few long-lived features. One of the first features *Voyager 2* resolved in far encounter shots was a large blue storm reminiscent of Jupiter's Great Red Spot. (But unlike Jupiter's cyclone, Neptune's spot disappeared within a few years.) The storm was approximately the same size relative to the planet, a deep pool floating in an azure sea of clouds. Scientists christened it the "Great Dark Spot," a reference to Jupiter's Great Red Spot. The storm spanned the distance of Earth's diameter.

Many of the exoplanets being discovered today resemble – either in mass or in location relative to their own star – the ice giants of our Solar System. As we look at Uranus and Neptune, we may be catching a glimpse of a class of planet that is common in our universe. It behooves us to study these frigid siblings for this, if for no other reason.

HOW DO THE ATMOSPHERES STACK UP?

As we have seen, the size of a planet affects a host of variables. The more mass a world has, the more gravity it holds. The bigger the core, the more likely it is to have a magnetosphere. Size also affects the amount of atmosphere a planet has. Some worlds are barren, airless places, while a rich brew of gassy chemistry blankets others. If a planet's core generates a magnetic field, this magnetic "bubble" will shield the atmosphere from solar wind, which can strip atmosphere away. More gravity tends to hold on to more atmosphere. Even temperature plays a role. Heat is merely the movement of atoms and molecules. If an atmosphere is heated, its molecules are in rapid motion, and can leak away into space more easily than a quiet atmosphere. Saturn's moon Titan has an extended atmosphere despite the fact that it has no magnetic field and low gravity. In size, Titan is practically a twin to Ganymede, an airless world. Why the difference? Part of the reason is that Titan's cold atmosphere clings to the small world in quiet stability, moving sluggishly in the moon's weak gravity. If Titan were as close to the Sun as Mars is, much of its atmosphere would be lost to space.

Fig. 4.14 Atmospheres of the planets reflect their sizes and proximity to the Sun. Here, we compare atmospheres scaled to pressure (Earth = 1 bar) but not to size. Notice how the cloud decks of Jupiter are compressed together compared to the other giants. This is because of its intense gravity (Art by the author.)

Because no surface exists on the gas and ice giants from which to measure altitude, scientists instead use pressure as a reference point, with 1 "bar" (equivalent to Earth's atmosphere at sea level) as the starting point. For purposes of comparison, our chart will delve down to 300 times that pressure, 300 bars.

Like the other spherical planets we've visited, the gas and ice giants are in hydrostatic equilibrium. This may seem strange for planets with no solid surfaces, but the atmospheres of the giants are self-supporting. The great pressures of their deep atmospheres act as a foundation against the force of gravity, keeping the atmospheres in balance with the pull of the planet. (We will see a similar situation with stars in Chap. 5.)

PLANETARY RINGS: ANOTHER SIZE CONTRAST

Planetary rings provide us with another useful scale for comparison. Although all the gas and ice giants have them, their structures vary from one planet to another. Planetary rings vary in size, consistency and brightness, ranging from Saturn's spectacular system to Neptune's incomplete ring "arcs." Scientists believe that the differences come from the fact that each ring system may be in a different stage of decay, with those at Uranus and Neptune being the most ancient. Rings and their gaps are also shaped by the gravity of various moons, depending on locations of the satellites' orbits.

Jupiter's rings, the consistency of cigarette smoke, are the least extensive of the four ring systems. They spread a nearly invisible tenuous halo around the planet. Jupiter's system holds three rings in tow. The outer edge of the main band begins abruptly at 78,000 km from the center of the planet. Its inner edge fades out gradually toward the planet at about 71,000 km and opens up into a great toroidal doughnut some 20,000 km thick. This is sometimes referred to as the Halo Ring. Outside of the main band is the Gossamer Ring. Its density thins out until the ring disappears at about the orbit of the tiny moon Thebe, 220,000 km out.

More extensive rings await us around the worlds further out. Saturn has been called "Lord of the Rings," and rightly so. Although Jupiter may be king, Saturn's rings make the planet's physical presence vast. In the quarter of a million mile gap between Earth and the Moon, Saturn's gargantuan ring system would span three-fifths the distance. The rings appear to be composed primarily of ice, although some portions may contain ice-covered rock. Spiral density waves move throughout the system like the grooves on an old LP record. Smaller moons, often embedded within the rings themselves, send elegant ripples, waves and gores across thousands of kilometers along the racetrack-like bands of ring particles, while 3.2-km-high chevrons of material rear up at the edge of the B ring. The gravity of small moons propogates these varied phenomena. The influences of larger moons, farther out, cause resonances with the particles, clearing gaps in the rings.

Fig. 4.15 Earth and Moon are seen here at their correct size and distance, with Saturn in between (Earth/Moon images courtesy NASA. Saturn image courtesy NASA/JPL/SSI.)

Fig. 4.16 Sixteen continental United States' could fit from the inner to the outer edges of Saturn's mighty ring system (Ring system photo courtesy NASA/JPL/SSI; art by the author.)

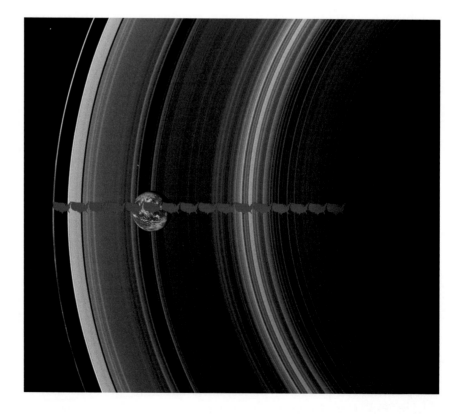

From the inside of the C ring to the outside edge of the ring system, sixteen continental United States would fit end to end. Despite their vast diameter, average ring thickness is equivalent to a three-story building. Put another way, if Saturn's rings were as thick as an audio CD, they would have a diameter of roughly 30 km. That's a thin CD.

A curious thing happens when we try to compare the scope of the four ring systems. If we shrink Jupiter and Saturn down far enough that their diameters match those of Uranus and Neptune, we can compare the rings visually. We immediately notice that the ring systems are all of similar extent. Those of Neptune seem to be on the small end of the scale, but as it is the most remote of the four worlds, more rings may be discovered with time. This was certainly the case with Uranus. Its faint rings were discovered in 1977 by ground-based telescopes, and two more were spotted in *Voyager 2* encounter images. But more discoveries awaited, as Hubble Space Telescope imagery revealed two new outer rings long after the *Voyager* flyby.

Saturn's rings are the most massive, and Jupiter's are the finest-grained, while the others lie in between. The rings of both Uranus and Neptune are black as coal. At Uranus, several amorphous, extended rings bracket a series of tightly bound inner rings. The inner rings have crisp edges, which may betray some moons yet undiscovered. Saturn's most similar rings, thin and ropelike, are all shepherded by small moons on either side. The well-defined Uranian rings consist of meter-wide black boulders. When *Voyager 2* looked back through the rings at the distant Sun 19.2 AU away, it discovered that the entire inner Uranian system is filled with fine dust, perhaps future rings waiting to form.

Neptune's ring system is chunky. Although most of its rings complete a circle all the way around the planet, many include thicker, arc-like regions. The outermost Adams Ring is the most dramatic example, with at least five contained arcs spotted by *Voyager*. The three largest have even been named: Fraternite, Egalite and Liberte. These ring arcs vary in extent, covering from 1 to 10 of the Adams Ring's overall length. Neptune's Leverrier and

Fig. 4.17 Shepherd moons herd Saturn's F-ring into a thin corridor. Note the gores and waves in the ring material (Image courtesy of NASA/JPL/SSI.)

Fig. 4.18 The four ring systems of the giant planets compared. Each planet's radius has been shifted to match the others, so that similarities and differences in the rings are more evident (Planet images modified from NASA/JPL images. Art by the author.)

Fig. 4.18 The four ring systems of the giant planets compared. Each planet's radius has been shifted to match the others, so that similarities and differences in the rings are more evident (Planet images modified from NASA/JPL images. Art by the author.)

Arago rings are narrow and concise, while Galle and Lassell are broader and fainter. Just inside the Adams Ring orbits a series of ring arcs, offering the appearance of an irregular dashed line. Its peculiar nature may arise from disturbances by the 150-km moon Galatea, which orbits just a thousand kilometers inside the Adams Ring. As Neptune is the farthest of the major worlds, it may well have more rings to be discovered, perhaps beyond the Adams Ring. If so, its ring system would bear even more similarity to those of the other giant planets.

The size and composition (mass) of a planet influence its gravity, which, in effect, determines how extensive its ring system may be. It appears that when we scale a planet down, its ring system scales down in similar fashion, as its gravitational influence shrinks with its mass. But we have also seen how size affects the very nature of a planet. We have witnessed a variety of scale within the planets, and that size diversity has resulted in a rich and varied planetary system. Far from cratered, dead spheres and simple globes of gas, the worlds of our Solar System exhibit wide ranges of environments, disparate atmospheres and temperatures, and an assortment of landscapes, cloudscapes and seascapes. From the terrestrial marbles to the gas giants, planets tend to exist as part of complex families, orbiting a central point. At that central point lies a star. But the stars represent a remarkable assortment of sizes, temperatures, colors and power. The stars are the next stop on our journey of scale.

Chapter 5
Bright, Shining Stars

Now that the scales of the moons and planets of our Solar System have become somewhat familiar to us, we can venture to much more distant objects, other suns. And as we saw with the planets and smaller bodies of our own Solar System, size determines a variety of factors. For stars, size determines how hot and bright they are, along with the course of their lives and their ultimate fate. In fact, size has a more profound effect on the nature of stars than it does on planets and moons. But before we visit the stars out there, we'll take a moment to get our bearings by visiting our own star, the Sun.

It's big. It's bad. It could hold as much material as makes up 1,300,000 Earths, without even loosening its belt. It's an important object: the Sun holds our entire planetary system – terrestrial and giant planets, asteroids, Kuiper Belt objects and Oort Cloud inhabitants – together in its mighty grasp. And although we measure the length of day by sunrise to sunset, even the Sun has days, turning once every 609 h, 7 min.

With the exception of nuclear energy, every form of power used by humans originates within the Sun. The fossil fuels (oil, gas and coal) we burn were once plant and animal matter. Those plants converted solar energy into sugars in the process of photosynthesis. Animals ate those plants, depending on the Sun's energy on a secondary level. Many cultures burn plant matter for energy. Others use solar power, directly accessing the Sun's energy for electricity. Still others use wind power, but our weather systems are driven by the Sun's heat. It all traces back to our nearest star.

Fig. 5.1 The red giant Betelgeuse dwarfs our Sun, at right (Art by the author.)

M. Carroll, *Picture This!: Grasping the Dimensions of Time and Space*,
DOI 10.1007/978-3-319-24907-0_5, © Springer International Publishing Switzerland 2016

The Sun radiates a steady flow of charged particles, the solar wind. This stream blows about 450 km a second throughout the Solar System. Occasionally, a barrage of particles explodes from the Sun in a solar flare, interrupting satellite communications and knocking out power grids on Earth. Flares usually erupt from sunspots, storms on the surface of the Sun. Sunspots appear as dark blemishes, although they are dark only in a relative sense (though still bright enough to instantaneously blind any wayward solar tourist). Sunspots mark cool regions where the Sun's magnetic field pierces the surface and loops out into space, carrying incandescent material with it. These gigantic moving streams, called prominences, tower dozens or even hundreds of times as tall as Earth's globe.

Deep in the Sun's core, nuclear fusion reactions burn hydrogen, converting it to helium. This generates an incredible amount of energy, far more than a simple chemical fire or gravitational collapse could yield. Photons – particles of light – transport this energy through the Sun's midshell, called the radiative zone, to the next layer of the star, the convection zone. There, mixing motions of gases bring the energy to the surface like boiling water brings heat to the surface in a pot of water. The journey of the photons takes more than a million years.

If the Sun is just a really big ball of gas with lots of gravity, why doesn't it collapse under its own weight? The answer lies in that balancing act called hydrostatic equilibrium, the same force responsible for the shape of planets. The Sun's gravity pulls its material toward the core, but the core generates tremendous pressure due to the fusion reactions going on inside. This pressure pushes out, balancing the gravitational inward pull and keeping the Sun in its spherical form.

Fig. 5.2 Twisted magnetic fields (seen here as bright blue lines) blast from sunspots (dark blue) on the Sun's surface (black). Iron atoms stream along the field lines, causing them to glow in ultraviolet light as they blow through the Sun's fiery maelstrom. Many of the lines loop back into other sunspots, linked by opposite magnetic polarity. A train of four Earths floats above the conflagration for scale (Sun image courtesy of NASA/Solar Dynamics Observatory; Earths courtesy NASA; art by the author.)

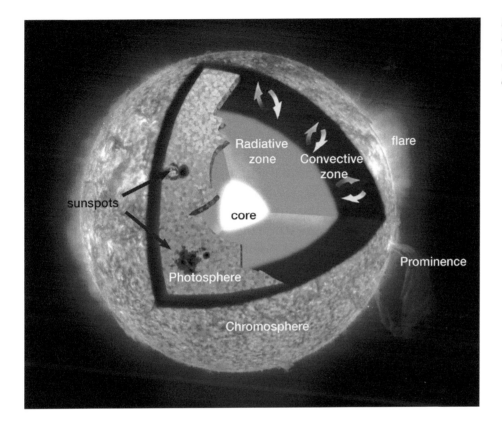

Fig. 5.3 Our Sun is arranged like the layers of an onion, with various zones nested within each other (Art by the author.)

LIFE CYCLE OF A STAR

The universe is alive with stars of many different sizes, colors and ages. The size of a star at its beginning will determine much about its makeup, the length of its life, and even the nature of its death. The life cycle of any star begins as a gigantic disk of gas coalesces into a growing sphere. As that sphere gains mass and gravity, it collapses in upon itself, breaking atoms down in its core and triggering nuclear fusion. Stars spend most of their time, roughly 90 % of their life spans, burning hydrogen into helium. This is a very stable time of life, where stars burn steadily and brighten over time. The energy of the hydrogen fusion "holds up" the star's outer layers above its core.

All is well…for a while. But as a star ages, the helium begins to fuse, outpacing the hydrogen fusion, and dramatic changes take place. Eventually, the hydrogen runs out, and what's left is a heavy shell of helium. The star's core begins to compress even more, abandoned by the supportive force of hydrogen fusion. The helium on the outside, along with leftover hydrogen, expands and heats up even further. The star grows larger and brighter, often transforming into a red giant. During this phase, our Sun will expand to fill the orbits of Mercury and Venus, perhaps even making it out as far as Earth's 1 AU. Stars the size of our Sun today also fuse heavier elements such as carbon. Larger stars generate more varied elements.

Finally, when there is no more fuel for the star to burn, it departs from what is called the "main sequence" and begins to die. This brings us to our earlier observation, that size determines the fate of a star. Medium- to low-mass stars like our Sun go fairly quietly into the night. They swell into a giant star and eject their shell of spent hydrogen and helium into space, creating a spherical cloud expanding around them. This globe of lustrous gas becomes one of the most beautiful phenomena in the universe, a planetary nebula (see Chap. 6). Left behind is a "dead" star called a white dwarf. The flash of brilliance concurrent with the star's expulsion of its gas shell is called a nova.

The larger stars buy the farm in much more dramatic ways. If a star weighs in at four to eight times the mass of the Sun, it will perish in a colossal explosion called a supernova. A typical supernova may become as luminous as all the stars in the galaxy, putting out as much power in an instant

Fig. 5.4 The famous Hertzspring-Russell diagram shows the life cycle of stars. The main sequence is the river of light flowing diagonally from upper left to lower right, with a few select stars labeled. As a star ages, it leaves this corridor and becomes a dwarf star (lower left) or a giant (upper right). The numbers next to each star show its mass, with 1 = the Sun. Notice how the mass of the star affects its brightness (scale at left) and its ultimate fate. The Sun will become a red giant like Betelgeuse. Larger stars such as Sirius will eventually become bigger giants such as Aldebaran, and the still larger Bellatrix will grow to blue supergiants similar to Rigel. Stars at the low, or cool, end of the main sequence live longest, while ones to the upper left burn brightly and die early. In this diagram, stars are not to scale (Diagram by the author.)

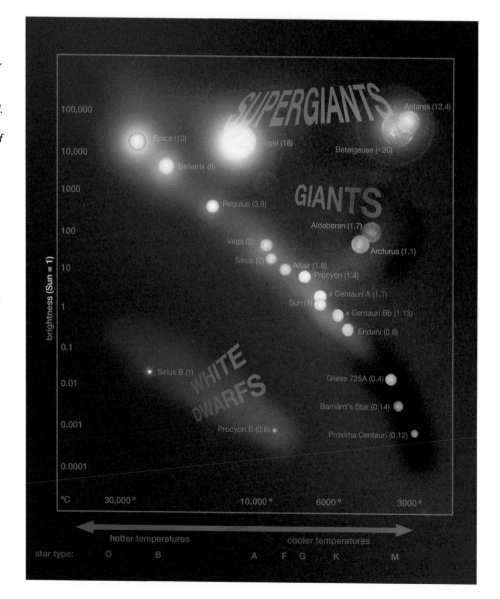

as an average star does during its entire lifetime. The resulting wave of gas may expand at up to 10 % of the speed of light. The material ejected from a dying star drifts into interstellar clouds throughout the galaxy, eventually coalescing into a new generation of stars. It's the circle of life, stellar-style.

Stars are classified into seven categories, according to their spectrum. Notice how their sizes affect their temperature, brilliance and even length of life. The more mass a star has, the brighter it is and the faster it burns its fuel. First-generation stars are all born with roughly the same makeup, mostly hydrogen and helium (about 98 %). In the early universe, all stars consisted of these two elements. But once they began to go nova or super-nova, the scraps of their explosive deaths resulted in heavier elements that combined to make later generation stars. These later stars, our Sun included, still have a majority of hydrogen and helium, but also contain heavier elements such as iron, lithium and calcium.

The laws of nature governing a star's outcome are fairly straightfor-ward and consistent. During its primary hydrogen fusion phase, a star's nature – its temperature and luminosity – are regulated primarily by that one all-important characteristic – its volume (and how much mass is con-tained within that given volume). Just as a realtor will tell you that the critical elements of real estate are "location, location, location," the critical elements of a star's nature are "size, size, size."

As we've seen, the size of a star also determines its longevity. If a star starts out huge, it will burn bright and hot, but not for long. Large suns burn through their fuel quickly. The star Spica, with a mass ten times that of the Sun, will live for about 10,000,000 (10 million) years. Our Sun will have a lifetime spanning the course of 10,000,000,000 (10 billion) years. The small, cool dwarf star Proxima Centauri will last perhaps 90 billion years longer.

Some nearby stars are quite exotic. Take, for example, the nearest star to our own Sun, Proxima Centauri. Proxima is part of a triple-star system.[1] It is a red dwarf star about half again the size of Jupiter (roughly 209,000 km in diameter). Proxima orbits two larger stars that are similar in size to our own Sun, Alpha Centauri A and Alpha Centauri B. It orbits a fifth of a light-year from the others, taking half a billion years to make the circuit. In its long, looping orbit outside of its companions, Proxima is now on the Earth-facing portion of its orbit, making it the closest star to Earth by a trillion kilometers.

Alpha Centauri A is the same stellar type as our Sun (G2), but it's a little larger. Surface temperatures reach 5500 °C (comparable to the Sun's), but its greater diameter, 25 % more than Sun, makes it 1.6 times as bril-liant. Alpha Centauri B is smaller and more orange. Its type is known as spectral type K2. Its lower temperature (5000 °C) results in the star giving off only half the luminosity of the Sun. These two primary stars circle each other once every 80 years at an average distance of 11 AU. The red dwarf Proxima Centauri simmers at just 2825 °C, and glows like a dying ember, only 1/500th as bright as our Sun. Proxima is known as a flare star; the surface exhibits sudden changes in brightness.

[1] Researchers are actually trying to confirm that Proxima is gravitationally tied to the two primary Alpha Centauri stars. It may simply be passing through the neighborhood.

Fig. 5.5 The red dwarf Proxima Centauri, nearest star to Earth, orbits two sun-like stars (at right). Humble Proxima will live hundreds of billions of years, while its companions, both with about the size of our Sun, will die in just a few billion years. Here, the system is seen from a hypothetical planet (Art by the author.)

How close is the Alpha Centauri system? Let's shrink our mighty Sun down to the size of a U. S.-regulation golf ball (42.67 mm diameter). This means that the golf ball Sun is 32.3 billion times as small as the real Sun! If we divide the distance to the Alpha Centauri star system by this number, we get a distance equivalent to the gap between Los Angeles and Albuquerque, or the distance from Madrid to Marrakech. At our golf ball scale, three normal paces will take us beyond the orbit of Neptune, and our journey to the nearest star has just begun. There is a lot of empty space in space. We look up into the night sky and see a spattering of jewel-like stars clumped together, but this is an optical illusion. The gaps between them are immense.

Proxima is among the smallest stars, called M-class stars. These little suns boast the longest lives. Known as red dwarfs, these dim, cool stars range from 0.075 up to half of the Sun's mass. Their surface temperature hovers around 3000 °C (the Sun reaches 15 million °C), yielding a dull red brightness of just three thousandths that of our Sun. Approximately 75 % of all stars in the universe are of this type.

Because of their slow-burning natures, these stars may have a lifetime of 600 billion years. While larger stars collapse after burning through the hydrogen in their cores, red dwarfs burn all of their hydrogen, from top to bottom, gradually. Then, like their larger cousins, they collapse into small white dwarfs. Nearly 20 out of the 30 stars nearest Earth are red dwarfs.

Next in line are K-type stars, sometimes referred to as orange dwarfs. These small suns can have as much as 0.8 solar masses. Slightly hotter than M stars (by about 1000 C), they shine at about three-tenths of the Sun's luminosity. This slow burn extends their lives, giving them a sell-by date of roughly 15–30 billion years. Such stars are fairly common, making up about 15 % of the main sequence population. Alpha Centauri B is a K-type star.

Our Sun is a G-type star. With its stable output and surface temperature of 6090 °C, its life expectancy is roughly 10 billion years. Feel free to invest in long-term stocks. G stars make up 7 % of all main sequence stars. They range in mass from 0.8 to about 1.2 solar masses. Alpha Centauri A is a G-type star.

F stars account for 2 % of the stellar populace. These stars span diameters up to 1.5 times that of our home star, with luminosities seven times as bright and temperatures of 6500 °C. Burning their fuel hotter and faster, these suns have a life expectancy of 2 billion years. One of them, Procyon A, is abnormally bright for its type, and it may be in the first stages of swelling to a red giant. While F-type stars account for just 2 % of all stars, the next stars on the list are even more rare.

Just 1 % of stars fall into the A class. These hot stars burn with a surface temperature of 8000 °C and a luminosity of 20 times that of the Sun. Weighing twice the mass, they burn out in roughly a billion years. The brightest star in Earth's sky, Sirius A, is one of these.

Rarer still are the B-class stars, known as blue giants. Out of a thousand stars, only one will be of this type. The searing temperatures of B stars reach 15,000 °C, burning with a thousand times the brightness of our own Sun. They contain between 2 and 16 solar masses. B-type stars rotate quickly, with equatorial speeds reaching 200 km/s. They produce fierce stellar winds of up to 3000 km/s. B stars pay the price for being so energetic. They will be cold, dead white dwarfs within 50 Ma of their birth.

Finally, the scarcest star of all is known as a blue supergiant. Having the spectral type O (toward the blue/violet end of the spectrum), these massive stars blaze with surface temperatures surpassing 50,000 °C, shining with a million times the light of the Sun. As they swell to the supergiant stage, their light may increase to several million times the brightness of the Sun. They can weigh in at between 15 and 90 solar masses. But like those sad cases in Hollywood, these bright celebrities burn out quickly, at the age of 500,000 years. Most of them are large enough to go supernova, leaving behind a rapidly spinning neutron star or even a black hole. The ancient central stars of many planetary nebulae such the Dumbell or Ring nebulae, are old O-type stars, formerly blue giants or supergiants. (For more on nebulae, see Chap. 6.)

A LOOK AT SUPERNOVAE

Supernovae fall into several specific classes. These exploding stars are separated by the chemical makeup shown by the starlight shining from them. The light spectra of Type I stars show no hydrogen, but hydrogen is present in Type II supernovae. Each of these families, in turn, has smaller clans within. A Type Ia supernova occurs when a white dwarf star pulls mass from a nearby companion star, increasing the dwarf's mass and leading to a thermal runaway effect. The temperature and mass continue to increase, and the core of the white dwarf finally ignites a type of fusion different

Fig. 5.6 The exploding star Eta Carinae ejects two great lobes of material. The star itself is 100 times the mass of our Sun, and its bipolar cloud of surrounding gas reaches 0.6 light-years across (Image courtesy of John Morse, STScI, Hubble Space Telescope.)

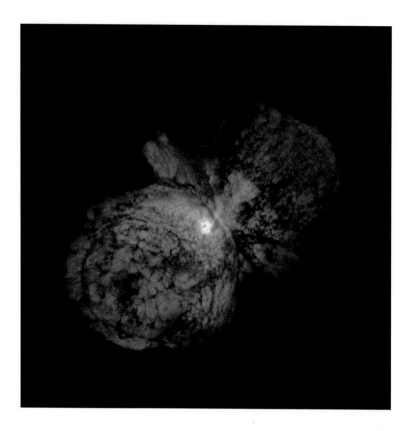

from the hydrogen-helium fusion of a normal star. When this new type of fusion – called carbon fusion – kicks in, the star ruptures, exploding with a luminosity 5 billion times its original brightness. Some Type I supernovae are also the result of the collision of two white dwarfs. Type Ib and Type Ic stars are related, but exhibit slightly different elements in their spectra.

Hydrogen-rich Type II supernovae fall into categories of Type II-P, Type II-L and Type IIn. The spectrum of their starlight is often smudged into wide bands, indicating that they are expanding at breakneck speed, sometimes up to thousands of kilometers per second. As in the latter cases of Type I's, all of these cases involve core collapse.

Some supernovae don't fit into either of these groupings. These "peculiar" stars are classified as Type III, IV or V. Observers 150 years ago witnessed the "great outburst" of Eta Carinae, a close star that brightened dramatically. The aftermath of Eta Carinae's violent outburst was imaged by the Hubble Space Telescope (see Fig. 5.6).

NEUTRON STARS

Supernova cores may collapse into super-dense neutron stars. A neutron star forms as the core continues to collapse until the entire star – often the mass of the Sun – is only 20 km in diameter. As the core's gravity draws in upon itself, protons and electrons combine to make neutrons, which is how neutron stars get their clever name. The gravity of an average neutron star is 2 billion times that of Earth. The explosion from the neutron star's collapse spins the core many times each second,[2] spewing out radiation like a lighthouse sends out beams of light. These pulses of radiation, usually X-rays and gamma rays, gave early radio astronomers a shock, says Dr. Dirk Terrell, astrophysicist and star expert at the Southwest Research Institute. "The discovery was these very regularly repeating radio signals, ticking away like an extremely accurate clock. Some people were shocked by this; there was no known natural phenomenon at the time that would lead to that sort of behavior. They were originally labeled LGMs, for 'little green men,' thinking they were some sort of extraterrestrial beacon, perhaps waypoints throughout the galaxy. It turned out to be rapidly rotating neutron stars."

Neutron stars generate savage magnetic fields that cause charged particles to move along magnetic field lines. As they do, they give off radiation. Most of that radiation beams out of the magnetic pole of the star, Terrell explains. "When the pole points toward Earth, we see it as radio waves. Then, as the neutron star points away, we see and hear nothing. Then it sweeps back around and we get another blip, so you get these repeated radio signals. The rule of thumb is that most stars start out close to the Sun in mass. When they shrink down to a white dwarf, they're roughly the size of Earth. But if you have enough mass, the collapse continues and you end up with a neutron star that's roughly the diameter of a city like Denver."

Since the demise of a star depends on the mass it has as it undergoes this collapse, what does it take to make a neutron star? Stars about the size of the Sun will end up as white dwarfs. The fusion in a star acts as a support. Electrons repel each other, forming a sort of force field in the core. This force is called electron degeneracy. Electron degeneracy works for stars up to about 1.44 solar masses.[3] But there is a limit to the force that even electrons can provide. If the star starts out with more mass than about 1.7 solar masses, "the fact that electrons don't want to get pushed together doesn't mean anything," Terrell says. "They get pushed together anyway. The electrons and protons shove together to form neutrons. Then it's the neutrons that prevent collapse. Basically, you end up with a giant ball of neutrons."

Some neutron stars are companions to much larger main sequence stars. Often, material from the nearby companion star will flow along intense magnetic field lines, spiraling into a disk of gas swirling around the smaller neutron star. As the gas is sucked up by the neutron star, it generates the waves of X-rays seen jetting out the poles, doing a good imitation of little green men.

[2] Some neutron stars have been clocked at 700 revolutions per second.

[3] This upper limit for the formation of a white dwarf is known as the Chandrasekhar limit.

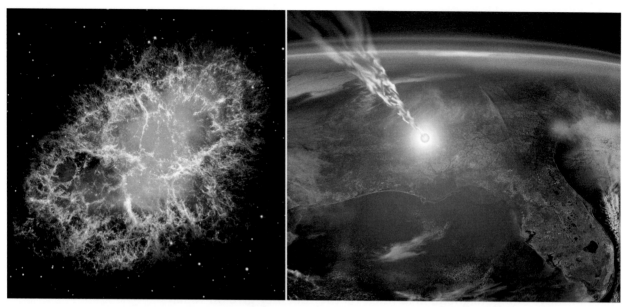

Fig. 5.7 Left: *The Crab Nebula is the leftover detritus of a supernova blast first spotted in A.D. 1054. Chinese and Islamic observers recorded the appearance of the "guest star," which was visible during the day. At the center of the expanding cloud is the collapsed core, now a neutron star. (Hubble Space Telescope image courtesy NASA/ESA.)* Right: *A typical neutron star is the size of a moderate city (Art by the author; background Florida modified from image courtesy NASA.)*

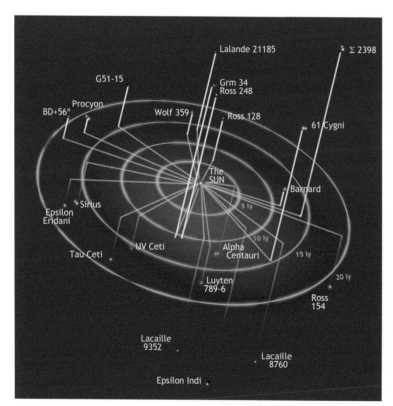

Fig. 5.8 *Our stellar neighborhood out to a distance of 20 light-years. The nearest stars, in the Alpha Centauri system, lie to the lower right of the Sun in this view (Diagram by the author.)*

BLACK HOLES

Ironically, the larger a star is when it leaves the main sequence, the smaller it becomes in the end. In the case of still larger stars than those that end up as neutron stars, Terrell explains, the cores collapse so far that they transform into a bizarre freak of nature. "If the collapsing core is greater than that limiting mass, it will blow right by the neutron star stage, and that's where we think you get a black hole. The weight of material overwhelms the neutron degeneracy pressure. As far as we know, there's nothing to stop it and it collapses to the singularity." This ultra-dense point allows no material – or even light – to escape. Astrophysicists detect them by the swirling material that falls into them, creating X-rays or gamma ray bursts. These collapsed stars are so dense that they bend the fabric of space/time around themselves in drastic ways. A star ten times as massive as our Sun becomes a black hole when it shrinks to the diameter of a city. Surrounding the black hole is a region called the "event horizon." At this boundary, time slows to a standstill.

STEPPING UP IN SCALE

We can gain insight into the nature of stars by visiting a few specific ones, carrying out our survey in a progression of size, and capping it off with a few bizarre outliers.

Although Proxima Centauri seems a runt compared to our Sun, several nearby stars are even smaller. Sirius B is a small companion to the brightest star in the sky: Sirius A.[4] Sirius A is a white main sequence star about 70 % larger than our Sun with twice the mass, but the star is hotter and 25 times as bright. It is also quite close to us, making it the brightest star in Earth's sky. Nearby Sirius B is small as stars go, slightly smaller than Earth. It is 10,000 times dimmer than Sirius A. The dim companion circles Sirius A in an elliptical 50-year orbit, swooping in as close as 8 AU and moving out to a distance of 31.5 AU, about the distance from Pluto to the Sun. As the leftover core of an exploded star, it is so dense that a piece of it the size of a sugar cube would weigh 998 kg. The white dwarf packs about .98 solar masses into its Earth-sized sphere, just under the Sun's mass. And while our Sun's surface simmers at about 5500 °C, Sirius B's surface is a searing 24,700 °C.

Like other white dwarfs, Sirius B started out burning a hydrogen shell that, in turn, converted into a helium-burning shell. Some 120 million years ago, at the beginning of Earth's Cretaceous period, Sirius B collapsed from a red giant into a white dwarf. The star suffered a titanic explosion that sent its helium crust out into space, converting some of it into metals that ended up in nearby Sirius A. It shrank into the tiny, diamond-like star we see today, a star that no longer burns with fusion.

[4] Although the stars of Alpha Centauri are closer to Earth, their combined light make them only the fourth brightest star in Earth's sky.

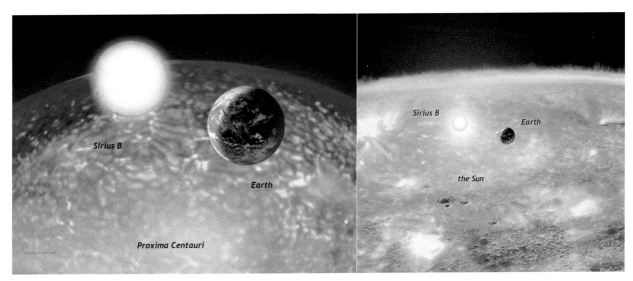

Fig. 5.9 Right: *The nearest star beyond the Sun, Proxima Centauri is about the size of Jupiter. Even this small star dominates Earth and the white dwarf Sirius B.* Left: *Our Sun compared to Earth and Sirius B (Art by the author.)*

Fig. 5.10 The blue supergiant Rigel looms behind Sirius and our Sun (Art by the author.)

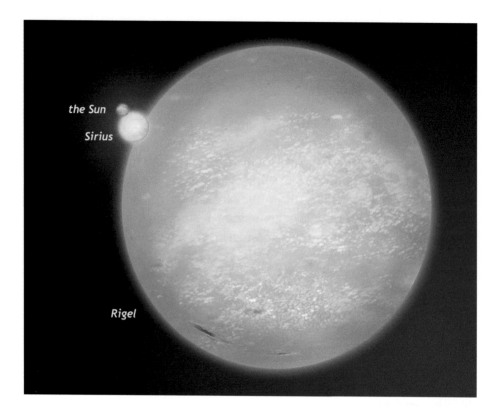

As we venture farther into the cosmos, we come to the star that forms the right foot of the constellation Orion, the Hunter. Rigel (or Beta Orionis), 850 light-years from us, is a blue-white supergiant 120,000 times as luminous and 21 times as massive as the Sun. Rigel is no longer a main sequence star. It has burned much of its primary hydrogen fuel and expanded into the hot behemoth that we see today. Rigel is not alone. Two nearby main-sequence stars wheel around it, circling each other every 9 days and 9 h. The larger of the two is 2.5 times the mass of the Sun, while the smaller is roughly 1.9 solar masses. These two stars, in turn, orbit the primary Rigel (Rigel A) on a lazy looping pathway 2200 AU away (12 light-days).

Rigel dwarfs the Sun, but larger stars are simmering out there. In the same constellation of Orion, a red giant makes up the hunter's left shoulder. Betelgeuse (alpha Orionis) is a titanic sphere of gas. Its light is variable, making measurements difficult. The latest estimates peg the star at 600 million miles across, or 400–600 times that of the Sun. If placed in the center of our Solar System, its gossamer edges would reach out to somewhere between the orbits of Mars and Jupiter. Put another way, if Earth was a golf ball, Betelgeuse would be as high as six Empire State Buildings.

Known as a red supergiant, the 10 million-year-old star is in the last stages of its life. Its surface has cooled to 3225 °C, providing it with its ruddy color. As it nears the end of its main sequence life, the star sheds a billowy envelope of gas 250 times the star's diameter. Betelgeuse travels at

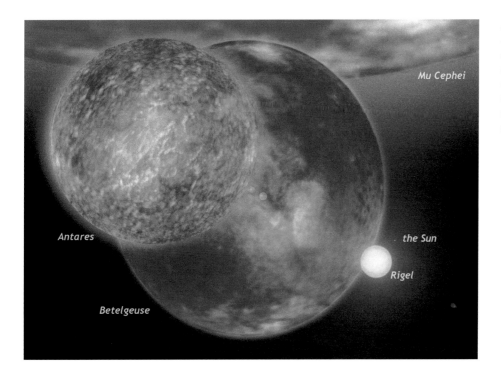

Fig. 5.11 The Sun floats as a small point to the right of the mighty supergiant Antares. Betelgeuse (using the lower estimate for its diameter) and Rigel are also shown to scale. Behind floats the behemoth Mu Cephei (Art by the author.)

30 km/s. Like a speedboat on a lake of interstellar gas, the star leaves behind it a bow shock 4 light-years across. Within the next million years, it will explode, becoming a supernova.

An even larger red supergiant glows at the heart of the scorpion, the constellation Scorpio. Although Antares spans a diameter 883 times that of the Sun, its density is one-millionth that of the Sun. Its fleecy surface smolders at a cool 3590 °C, just over half as hot as the Sun.

Three thousand light-years away is the red supergiant Mu Cephei, one of the largest stars known. Its diameter reaches some 1035 times as far across as the Sun. Compared to our golf ball Earth, Mu Cephei's diameter would stretch the length of two Golden Gate bridges end to end. Its remarkable color, noted by astronomer William Herschel early in the nineteenth century, is a deep crimson. The star is now referred to as "Herschel's Garnet Star."

NML Cygni is still bigger. Referred to as a red hypergiant, the star's diameter must be measured in astronomical units.[5] With a span of 15.3 AU, MNL Cygni would not fit within the orbit of Saturn. The enormous star shines with a power 560,000 times that of the Sun. With our Earth as the size of a golf ball, NML Cygni would stand slightly taller than Mt. Everest, tallest mountain on the planet.

Most stars are linked to companion stars. "It can get pretty complicated," says Dirk Terrell. "If we look at a triple system, the influence of that third star can be profound. It may cause the close binary to spiral in closer together. Binaries that have orbital periods of a few days or less are often members of triple systems, and it is the third star that enables the close binaries to spiral in to these close orbits we see."

[5] As measured by the European Southern Observatory's Very Large Telescope and Hawaii's Very Long Baseline Array; see *The Distance and Size of the Red Hypergiant NML Cyg from VLBA and VLA astrometry*, Zang, et al, arXiv: 1207.1850 [astro-ph.SR].

That fact has some interesting implications for the formation of exo-planets. A third star may inhibit planetary formation around close binaries. The Kepler observatory data hints at this. No planets have yet been found orbiting close binaries.

PLANETS OF OTHER STARS

While the Sun is a typical star, its family appears to be unique among the many hundreds of systems charted so far. "Our Solar System, it's now safe to say, is weird," says Caltech astronomer Mike Brown. "The majority of stars that we've looked at have a sort of Earth or twice-mass-Earth planets close to their stars. Some have hot Jupiters, big rocky or small gassy planets really close in. What we've never really found is something like the Solar System. The most common types of planets being discovered are these things between 2 and 5 Earth masses. We don't even have anything like that in our Solar System, so we don't have the most common type of planet, we don't have the most common locations of planets. We're seeing a lot of systems, and it's really clear that most planetary systems are just not like us."

Judging from the size of planets in our own Solar System, we expect to find far more smallish, terrestrial-sized worlds around other stars than gas or ice giants. Expecting them is one thing, but finding them is another matter. Several techniques have met with great success in the field of exo-planet searches. Perhaps the most straightforward is the transit technique. When an exoplanet passes in front of its host star, as seen from the viewpoint of Earth, a transit takes place. Though these stars are too far to actually see as a disk, light-sensing instruments can detect a drop in the star's overall light as the small planet eclipses it.

A limitation of this approach is that the planetary system must be seen nearly edge-on so that planets pass between their star and us. Astronomers have had practice with this phenomenon closer to home. Both Mercury and Venus sometimes pass directly between Earth and the Sun, blocking out a tiny spot on the Sun's disk. Transits do not happen in every orbit, as Earth and the other planets do not lie in exactly in the same plane. Venus transits occur only once each century.

What would we see from Earth if we took Jupiter, largest planet in our Solar System, and placed it in an orbit around a distant Sun-like star? Jupiter's diameter is about one tenth of our hypothetical star's, which means that its silhouette against the star's disk would cover 1/100 of the star's total surface. The level of sunlight reaching us would drop by 1 % as Jupiter passes across its disk. A planet the size of Earth is far smaller than Jupiter. The starlight from a transit of a small planet like Earth would drop by only 0.01 %. Astronomers have been able to measure a 1 % decrease in a star's light since the advent of photometry, a technology that arose in the 1950s. Today's instruments are more sensitive, so observers can regularly chart transits by planets even as tiny as Earth. Astronomers carefully watch the light of a star as it drops off. From this simple method, they can estimate the speed and orbital size of a planet. These characteristics also tell us something about how massive the planet is. But to confirm the existence of an exoplanet, the change in light levels must be observed multiple times.

The exoplanet-hunting Kepler space telescope uses the transit technique. Kepler is designed to detect incredibly minute differences in the location and movements of the stars it studies. To get an idea of just how accurate the observatory is, if the spacecraft observed twin streetlights on a street corner 1 km away, it could detect the movement of one of the lampposts by just 1 cm.

Another method for discovering exoplanets involves the tug that those planets have on their parent star. A planet causes the star it orbits to wobble. Jupiter causes our own Sun to wobble around a position outside its own radius with a period of about 12 years, the same time it takes the mighty planet to make one circuit. The larger the planet, the larger its star's wobble becomes. The other planets have more subtle effects, sometimes overlapping each other.[6] The problem is that the wobbling of a star is nearly indiscernible from a distance.

Rather than trying to detect the side-to-side swing of a star, it is actually easier to detect the movement of a star toward us or away from us using its light. Like sound, light travels in waves. Red light has longer waves than blue light does. Like a passing emergency vehicle with its siren blaring, the pitch of the siren's sound seems to rise on approach and decrease as the vehicle passes. This is because the sound compresses into shorter waves as the car approaches but stretches out as it departs. The effect is called the Doppler effect. Like a moving siren, objects traveling in space send out waves, but rather than sound waves, these are waves of light. If a star is approaching us, its light waves will be piled up

[6] In a similar way, while the Moon orbits Earth, it pulls on our world significantly. Seen from the outside, Earth "wobbles" by about 4600 km.

Fig. 5.12 The Kepler observatory uses the transit technique to sense exoplanets. Here, Venus crosses in front of the Sun in a transit seen from Earth (Image courtesy of NASA/SDI, AIA/Goddard Space Flight Center.)

and shortened, shifting toward the blue end of the spectrum. The light of objects racing away will be tinted red. Nearby stars move toward or away from us as they are tugged upon by nearby planets. This subtle shift in light is measureable to a fine degree.

MAJOR EXOPLANET DISCOVERIES

Canadian astronomers announced the first discovery of an exoplanet in 1988. The planet orbits the star Gamma Cephei, part of a binary star system. Because of the two suns in its sky, conditions on the planet are probably quite alien. But in 1995, observers detected the first planet in orbit around a Sun-like star. The star, 51 Pegasi, lies 51 light-years away. Planet 51 Pegasi b (informally known as Bellerofon) is a type of planet called a hot Jupiter, with a mass at least half that of Jupiter in a very tight orbit. Bellerofon is so close to its sun that surface temperatures hover around 1200 °C.

Since the discovery of 51 Pegasi b, the detection of exoplanets continued to come in, first at a trickle, and then in a flood. Even before the revolutionary Kepler space telescope planet-hunter, researchers were reporting ten new planets each month. In 2001, observers spotted the first planet within a star's habitable zone, a region where liquid water can remain on the surface.[7] The planet weighs in with a mass nearly 2000 times that of Earth, so it is clearly not an Earth-like world.

Kepler ushered in a renaissance of exoplanet research. Launched in 2009, the telescope's sole aim was to chart the light levels of 145,000 stars to detect the presence of planets. As it does, the craft is able to detect slight variations in light from planetary transits. With careful timing of the planet's orbit, scientists can determine the planet's size and mass, along with its orbital characteristics.

With some 6050 candidate planets under their belts, astronomers can now give us an overview of the general types of planets discovered so far. Several types make up the majority. Because of the relative ease of finding the largest, the list is skewed toward the big planets. First to be detected were the hot Jupiters, gas giants whose mass exceeds half that of Jupiter. Nearly half of the exoplanets with known masses fall into this category. Most have orbits dramatically unlike that of Jupiter. These monsters orbit their suns in close proximity, often more closely than Mercury orbits the Sun.

Because of their proximity to their suns, these worlds endure such high temperatures – up to 3600 °C – that some may grow a comet-like tail as their atmospheres stream off into space, stripped away by the nearby parent star. One such planet, cataloged as SWEEPS-10, careens around its sun so closely that it makes an entire circuit every 10 h. Some hot giants are close to Jupiter in size and mass, while others span a diameter more similar to Neptune.

[7] For our Solar System, Earth is in roughly the middle of our Sun's habitable zone, while Mars is just on the outside edge of it, and Venus is on the inner edge. If Venus had a thinner atmosphere, it could support liquid water on its surface. Temperatures on Mars reach a point at which liquid water is possible. Streaks of briny moisture have been spotted from orbit, wicking through the sand and migrating downward on some slopes.

Several hot Jupiters contain less than 20 % the mass of Jupiter. Two thirds of them weigh in at between a half and 20 Jupiter masses. One recent addition to the exoplanet menagerie lies just 30 light-years away, at the faint red dwarf star GJ436. Circling this star, the Neptune-sized planet GJ436b floats in a gargantuan cloud of hydrogen. The red dwarf's solar wind forces hydrogen from the gas planet's atmosphere, creating a comet-like tail some 50 times the size of the star itself. The hot world orbits less than 5 million km from its star – 1/30th the distance from Earth to the Sun – hurtling around it in just 2.6 Earth days.

Related to these searing worlds are the hot Neptunes. Like hot Jupiters, these planets circle close to their host stars (less than 1 AU). Unlike the frigid Uranus and Neptune we are familiar with, these giants are molten, holding 3–20 % of Jupiter's mass. Neptune-sized planets are the most common found so far.

Another frequent discovery has been that of cold Jupiter exoplanets. Cold Jupiters orbit at least 2 AU from the star, and are probably quite similar to Jupiter and Saturn in nature. Much more massive are the super-Jupiters, gas worlds with at least five times the bulk of Jupiter. Most have been found in distant orbits, but some are "hot super-Jupiters," orbiting close in. A recent example is the super-Jupiter Kappa Andromedae b, a giant planet orbiting the star Kappa Andromedae. Its mass is a whopping 13 times that of Jupiter. The planet is so large that it is similar to a brown dwarf, a globe nearly large enough to generate nuclear fusion, slightly smaller than an active star.

Super Earths constitute another fairly common exoplanet type. One out of ten exoplanet discoveries to date fall into this class. These worlds have no analog in our Solar System. The dense worlds are most probably similar in makeup to our terrestrial planets, with masses 1.5–10 times that of Earth. In 2007, Swiss researchers announced the detection of a super Earth orbiting the star Gliese 581, a red dwarf 20 light-years from Earth. In fact, three super Earths may orbit this star. Gliese 581c orbits at the inner edge of the star's "habitable zone," and may be Venus-like, suffering a runaway greenhouse effect. The other two, Gliese 581d and Gliese 581 g, may orbit further within the habitable zone, making them candidates for life.[8]

Just how "Earth-like" is a super Earth? Several features contribute to the uniqueness of our own world. First, Earth orbits in the Sun's habitable zone. Some super Earths undoubtedly orbit at such a distance from their own stars. Additionally, our world is large enough to have a molten core that generates a protective magnetic field around us. Many super Earths probably have such an energy field surrounding them. But another phenomenon sets Earth apart from other planets – plate tectonics (see Chap. 4). Earth's plates play an important part in our ecosystem. Recent work suggests that super Earths may not enjoy the benefits of plate tectonics. A team of researchers simulated the pressures within giant Earths, and found that the crust of such worlds is probably thick, preventing the movement of plates. Models suggest that within super Earth interiors, magma plumes – the kind that give rise to Earth's life-giving volcanoes –

[8] Some recent research has called into question the existence of the two planets, so it will take further research to determine what is going on at Gliese 581.

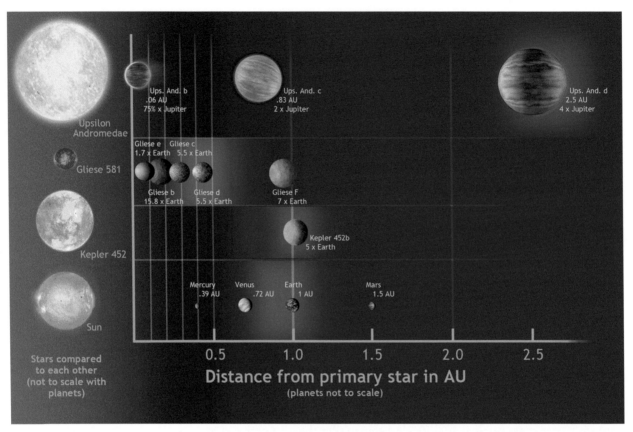

Fig. 5.13 Several exoplanetary systems compared in distance to their primary star. Planets are not to scale. Note the differences in the stars at left. These are correctly scaled to each other (but not to their planets). Planets Upsilon Andromedae d, Gliese c and d, and Kepler 452b (a super Earth) are in their star's approximate habitable zones, indicated in the glow behind (Art by the author.)

may stall out as they rise toward the crust. But some researchers contend that the increased heat within super Earths might be enough to drive plate tectonics after all.

Smaller on the scale of distant worlds are the exo Earths, rocky planets similar in size and mass to Earth. Their orbits vary widely, creating surface conditions ranging from hellish to possibly life-sustaining. With the Kepler data in hand, astronomers now estimate that there could be as many as 40 billion Earth-sized planets orbiting stars within habitable zones.

The closest match to an Earth-like planet is Kepler-452b, the first Earth-size planet to be found in the habitable zone of a star similar to the Sun. The planet is probably half again as large as Earth. Although Kepler-452b is slightly farther from its star than Earth is from the Sun, its star – Kepler 425 – is brighter, so the planet gets about the same amount of energy from its star as Earth does from our own Sun. Kepler-452b's atmosphere is probably thicker than Earth's. Because of its size and the likelihood that it is a rocky world, the planet may well have active volcanoes. It takes 385 days for the Kepler-452b to orbit its star, a year quite similar to Earth's 365-day year).

A final planetary type, outside the purview of Kepler, is called a rogue planet. Rogue planets must be on the order of Neptune-sized or larger to

Fig. 5.14 From left to right: *Jupiter, the hot Jupiter SWEEPS 10, the hot Neptune GJ436b, super Earth Kepler 581c, the Earth-like Kepler-452b, and our Earth (Jupiter and Earth modified from NASA/JPL photos; art by the author.)*

even be detected. That is, they must be large enough to warp the space around them in such a way that Earth observers can see distorted images of stars. This warping of space, called gravity lensing, bends images behind a heavy object (see Chap. 6). But the odds of finding a planet in this way are remote. Rogue planets are free-floating, wandering the cold cosmos between stars. In studies of some 50 million stars over the course of several years, researchers have tracked down 474 gravitational lensing locations. Of those, ten were consistent with a Jupiter-mass object. Researchers estimate that at least 75 % of the observed objects were not in orbit around a star. Instead, they are rogues, drifting aimlessly through the starry darkness of interstellar space.

MOONS AND LIFE

Because of the limitations of our search techniques, the data tends to be skewed toward larger planets, worlds very different from our own. Many range in size from Neptune-like to Jupiters on steroids. But smaller worlds may be linked to them. All four of the giant worlds in our own Solar System host families of moons. Some are as complex – and nearly as large – as terrestrial planets.

A recent study[9] asserts that, "With most known planets in the stellar [habitable zone] being gas giants between the sizes of Neptune and Jupiter rather than terrestrial planets, the moons of giant planets could actually be the most numerous population of habitable worlds." Just ask James Cameron; his *Avatar* took place on such a moon, orbiting a Neptune-like world in a habitable zone.

[9]Heller, R., et al. 2014. "Formation, Habitability, and Detection of Extrasolar Moons." *AsBio*, 14, pp. 798–835.

What kinds of moons can we imagine in the planetary systems of other stars? Statistics from our own Solar System suggest that swarms of moons circle the giant exoplanets the size of Neptune or larger. Since many of those host planets circumnavigate their stars in Earthlike orbits, conditions on any water-enshrouded moon would resemble those on Earth. But aside from the habitable zone, other factors come into play when thinking about life-friendly moons out there.

A myriad of processes probably influence habitability. For example, icy moons may well be subject to the same forces of tidal friction that heat Io, Europa and Enceladus. Geological activity within those moons may lead to the kind of subsurface oceans we've seen in our own system. However, exomoon expert Rory Barnes of the University of Washington's Astrobiology Department warns, "There could be processes that preclude moons from hosting life. But certainly if we assume you need a solid surface and liquid water, and we now know there's a lot of planets out there, there's a greater possibility of finding a lot more [habitable moons] than there are planets."

The scale of these exoplanets and moons has a bearing on the search for extraterrestrial life, Barnes asserts. "Size definitely matters. We want to know how big and how massive any body is before we assess it for habitability." Big planets tend to have big magnetospheres, the protective field that shields a world from incoming radiation. The larger a planet's active core, the more extensive will be this magnetic "bubble." In our own Solar System, the small terrestrial worlds have relatively weak magnetic fields (those of Mars and Venus are negligible), while ferocious magnetospheres enshroud the giants Jupiter, Saturn, Uranus and Neptune, with mighty Jupiter having the greatest and Uranus the least, corresponding to their sizes. Their massive cores, each the size of a terrestrial planet, are able to generate stronger fields.

A moon orbiting near a giant planet might actually benefit from its magnetosphere, as it will benefit from other features of the nearby world it orbits. "The planet provides some light and other energy to the moon," Barnes suggests. "Jupiter, for example, emits twice as much energy as it gets from the sun. Nearby moons could receive some energy from their host planet. That changes the climates and change where best habitats on a moon can be." As an example, Barnes cites Jupiter's icy moon Europa. Scientists have posited that high-energy particles hitting the Europan surface could provide food for a biome below. As Jupiter's radiation generates molecules on and within the ice, these materials seep down to the underground ocean. Radiation-produced molecules could then become fuel for marine organisms within Europa's subsurface seas, protected by the moon's ice crust. If the gravity of other moons and the parent planet are just right, tidal action might also generate volcanism on the seafloor of icy moons (or their watery counterparts closer in toward the star). This hydrothermal energy may also contribute to biology, as it does on our own planet. "It stands to reason that there will be possible habitats on some of those moons," Barnes says.

The size and sheer power of the stars around us boggle the mind. As Pastor Louie Giglio eloquently puts it, "It is not just 'twinkle, twinkle little star, how I wonder what you are.' I'm telling you what you are: intense and huge and massive and ferocious."[10] Although not couched in technical jargon, Giglio has a reasonable grasp of the nature of the many and varied stars scattered across the universe, from white dwarfs to red hypergiants to black holes. All of these stars came from clouds of gas floating through interstellar space. These nebulae, along with the star-island galaxies that host them, await us at our next stop up the scale.

[10] From the DVD *How Great is Our God* with Louie Giglio; Six Step Records/Capitol, August 2012.

Chapter 6
Nebulae, Galaxies and the Edge of All Things

Fig. 6.1 The incandescent "star-clouds" of the Lagoon Nebula drift in the solar winds of newborn stars, some only a few million years old. As seen through the eyes of the Hubble Space Telescope, this detail section of the nebula is 3 light-years across (Image courtesy of STScI.)

The gas from dying stars, a funereal celebration on a truly grand scale, results in some of the most stunning vistas in the universe. When stars leave the main sequence and enter old age, many end in dramatic explosions. Great waves of gas many times the size of a respectable solar system fan out into surrounding space. These clouds of drifting interstellar vapor, glowing in the starlight of many suns, are called nebulae.

Before we venture into deep space to survey these spectacular clouds and the galaxies that play home to them, we need to get our bearings once again. The scales we will cover next encompass a vast territory, but we'll watch for some similar forms that will help us find our way.

RINGS BEYOND THE PLANETS

Rings are a sort of universal form found on many scales – and for many reasons – throughout the universe (see the Preface). We will hopscotch across the universe in search of these forms, starting out small (at least on the cosmic scale). We begin our progression from the planet Jupiter, not in its rings but in its clouds.

M. Carroll, *Picture This!: Grasping the Dimensions of Time and Space*,
DOI 10.1007/978-3-319-24907-0_6, © Springer International Publishing Switzerland 2016

In 1993, Comet Shoemaker-Levy 9 (SL-9[1]) abruptly appeared on the scene, as comets often do. The difference between SL-9 and other comets was that it was not orbiting the Sun but rather orbiting Jupiter. The comet had apparently been captured on close approach, probably snagged by the combined gravity of Jupiter and its Galilean satellites. By the time astronomers spotted the comet, SL-9 may have been in orbit for as long as 27 years. Calculations suggest that the comet probably broke up during its initial close pass. Observers could instantly tell that something atypical was going on with the cosmic visitor. The comet consisted of at least 22 nuclei, arranged in a long row, each with its own tail contributing to the whole. But the comet's trajectory told an even more intriguing story: it would impact into the clouds of Jupiter the very next year, in July of 1994. Frustratingly, the impacts would occur on Jupiter's unseen far side in its southern hemisphere. Telescopes the world over came online to observe the aftermath of the string of impacts. Researchers also tasked several spacecraft with observing the events, including the Hubble Space Telescope and the Jupiter-bound Galileo, which would be in a position to directly witness the impacts from a distance.

The impacts surpassed projections. The first explosion belched material 3000 km into space above Jupiter. The fireball became visible from Earth, rising over the edge of the planet and fanning out like a nuclear mushroom cloud. Just minutes later, the impact site itself rotated into view, displaying a vast ring of dust the size of Earth. The train of dark spots appearing over the next 6 days marked debris from the impacts. The cometary fireballs triggered waves that moved through Jupiter's colorful clouds at 450 m/s. The largest impact, from "Fragment G,"[2] left a blemish 12,000 km across. Its enormous explosion released 600 times the energy of the world's entire current nuclear weapons inventory. Fully a

[1] Named after discoverers Eugene and Carolyn Shoemaker and David Levy.
[2] Fragments were labeled "A" through "W."

Fig. 6.2 *A succession of rings, from small to large, demonstrates echoes of form. Rings become progressively larger as images move clockwise from upper left.* **1.** *Comet Shoemaker Levy 9 deposited great rings of dust during its violent impacts in the clouds of Jupiter (Image courtesy of STScI, HST.)* **2.** *The rings of Saturn are the most well developed rings in our Solar System (Image courtesy of NASA/JPL/SSI.)* **3.** *Rings of gas and dust encircle the Sun-like star HL Tauri, making a disk-shaped cloud three times the size of our primary planetary system out to Neptune's orbit (24 billion km) (Image courtesy of Atacama Large Millimeter/submillimeter Array, or ALMA.)* **4.** *Rings of material come in many forms, including this odd spiral-armed disk around the star SAO 206462, 100 AU across (Image courtesy of Subaru Telescope.)* **5.** *The innermost disk surrounding the nearby star Fomalhaut spans just 0.1 AU, but the outer edge of the entire cloud is some 133 AU in diameter (Image courtesy of STScI, HST.)* **6.** *The Fine Ring Nebula is a great torus of glowing gas a third of a light-year – 21,000 AU – from rim to rim (Image courtesy of the European Southern Observatory.)* **7.** *Supernova 1987A lies in the nearby Large Magellanic Cloud dwarf galaxy and wears a necklace of ejected material 1 light-year across, the remnants of the star's death explosion (Image courtesy of STScI, HST.)* **8.** *When the dying star SGR1900 + 14 blew up, its blast tossed out a ring of dust some 3 × 7 light-years across (Image courtesy of NASA/Spitzer Telescope.)* **9.** *SNR 0509, a supernova remnant in the Large Magellanic Cloud, 23 light years across (Image courtesy of NASA/ESA Hubble Heritage team.)* **10.** *The elegant galaxy with an inelegant name, NGC7049, hosts a great ring of lacy dust 150,000 light-years in diameter (Image courtesy of NASA, ESA, and W. Harris, McMaster University, Ontario, Canada.)*

month after impact, the dust rings had faded dramatically. Eighteen years later, when the Pluto-bound New Horizons spacecraft flew by Jupiter, it charted disturbances in Jupiter's ring system, residual effects of the SL-9 collisions.

Planetary scientists have discovered evidence of similar multi-nucleus comet impacts near Jupiter. On the moons Ganymede and Callisto, crater chains called catena parade across icy landscapes. They mark trails of adjacent impacts, much like Jupiter suffered at the hands of Shoemaker Levy-9.

The planet-sized dust rings have faded from Jupiter's cloudy skies, but its ring system, along with those of the other giants, will remain intact for eons to come. These ring systems have much to tell us. Saturn's rings – the most extensive of all – provide insight into the dynamics of protoplanetary disks, much larger structures where entire planetary systems form. And beyond those dusty rings, we see similar structures in nebulae and even entire galaxies.

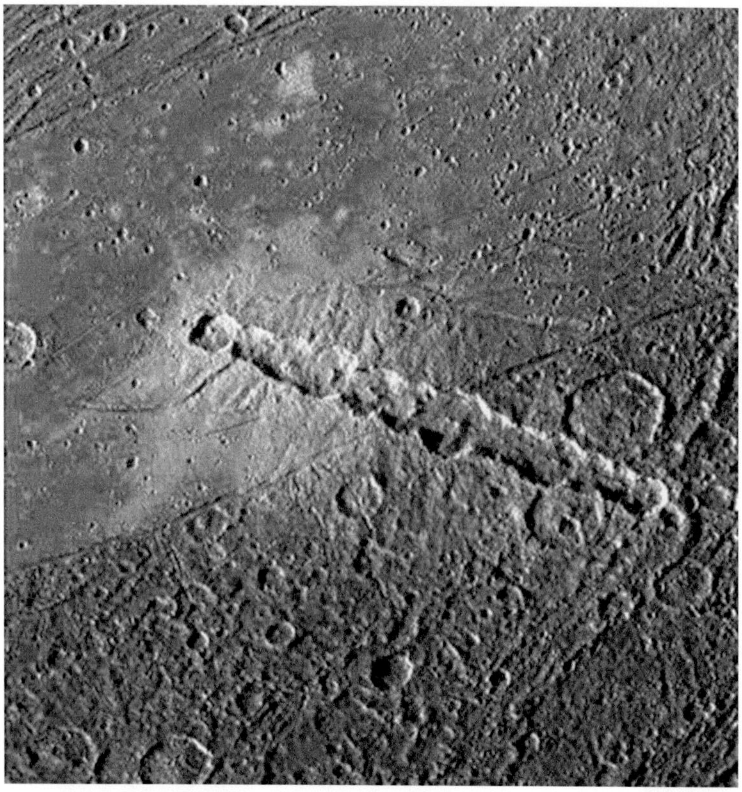

As we venture beyond our own Solar System, we come to a great cloud surrounding the young, Sun-like star HL Tauri, some 450 light years from Earth. Its Frisbee of extended material spans an area three times the distance from our Sun to Neptune (2,798,000,000 miles, or 90 AU).

Observations intimate the presence of at least one infant planet in orbit around HL Tauri, but there are undoubtedly more. Chile's Atacama Large Millimeter/submillimeter Array took images that may show the birth of planets, indicated in part by the dark gaps within the dust disk.

Fig. 6.3 Crater chains like Enki Catena on Ganymede evidence the impact of a fragmented body similar to Shoemaker Levy-9. Planetary scientists have confirmed 13 crater chains on Callisto and three on Ganymede (Galileo spacecraft image courtesy NASA/JPL/Cal-tech)

Although the dust rings around HL Tauri form roughly concentric circles, some disks are not so well-behaved. The star SAO 206462, imaged by Japan's Subaru telescope on Mauna Kea, Hawaii, cocoons itself at the center of a spiral-armed disk 100 AU across. If SAO 206462's dust ring does contain planets, their gravity and orbital motion could alter its form, tossing out its unique spiral arms.

A larger disk surrounds the nearby star Fomalhaut. Fomalhaut is a young[3] type A star approximately 25 light-years from Earth. Though only 1.84 times the size of our Sun, it is bright, blazing with light 16.6 times as powerful. An inner disk huddles within 0.1 AU (9.3 million miles) of the star. The disk seems to consist of tiny, carbon-rich motes of ash. The next ring out is a disk of larger particles with inner edge 0.4–1 AU of the star. Beyond that, an outer torus-shaped disk – Fomalhaut's "Kuiper Belt" – spreads to a diameter of 266 AU.

A thousand times larger, and expanding through space some 4900 light-years away, is the Fine Ring Nebula. The lovely blue-purple ring of gas marks the telltale relic of a star that shed much of its outer shell some 8700 years ago. Burning at the center of the Fine Ring Nebula is a binary star system.

Our scale ramps up as we visit a dead star in a galaxy far, far away. In 1987, a supernova erupted in the Large Magellanic Cloud, a satellite galaxy of the Milky Way some 168,000 light-years away. The detonation of Supernova 1987A left a double loop of material above and below it, along with an inner circle of expanding gas and dust. That inner ring is a full light-year across. Waves of star-stuff collide with the ring, causing 30–40 pearl-like "hot spots" to glow. These knots will likely combine to form a continuous, incandescent circle in the near future.

Still larger – and a bit closer to home – is a bizarre ring of material surrounding the remains of SGR 1900 + 14, a star that committed stellar suicide 20,000 years ago. The dead star at the center is a strange object called a magnetar (for magnetic star). Magnetars are the remains of the cores of massive stars that blew up in supernova explosions. What sets them apart from other dead stars is their fierce magnetic fields. Astronomers theorize that a cloud of dust enshrouded SGR 1900 + 14, and its explosion blossomed through the cloud, pushing out the dusty ring we see today. The ring is elliptical, traversing about 7 by 3 light-years.

Next on our agenda of expanding scale is another supernova in the Large Magellanic Cloud, this one some 23 light-years across. The white dwarf star SNR 0509 blew up 400 years ago, leaving a flimsy shell of material drifting through space. The giant bubble formed as a surge of gas slashed through the nearby interstellar clouds. The shockwave is racing through space at a speed of over 18 million km/h. Astrophysicists believe that the white dwarf was part of a binary system. It pulled material from its companion star, taking on so much mass that it became unstable, finally shattering.

Our voyage of increasing size brings us to NGC7049. This galaxy covers 150,000 light-years in diameter. The galaxy is surrounded by a remarkable rope-like ring of dark dust that stands out against the light of billions of stars behind it. The galaxy's bizarre structure may have resulted from a recent collision of two galaxies.

---

[3] A 2012 study estimated that Fomalhaut is only 440 million years old.

NEBULAE: THE NEXT STOP OUT

The glowing mists between the stars exhibit many different colors, for various reasons. Some nebulae simply reflect the light of nearby stars. Other clouds are energized by the power of stars within. Neighboring stars ionize their gases, causing them to glow in spectacular colors against the velvet black of space. Blue clouds come from starlight reflected upon dust grains, the cinders of dead stars. Great dust lanes block our view, creating ominous silhouettes against the starlight behind. These clouds can take on fanciful forms, and have inspired names such as the Horsehead, Cone, or Pelican nebulae.

Stars pour their material back into the space in one of two ways. Stable stars like our Sun produce a steady stream of stellar wind. Mid-sized suns become red giants. As they do, they blow off their outer layers into a sphere called a planetary nebula. Still larger stars don't last as long, exploding as supernovae. Gas ejected into space by these suicidal stars creates a bubble of hot gas, and leaves behind a tiny burned-out core as a neutron star or even a black hole.

DETERMINING THE SIZE OF COSMIC STUFF

In order to understand how large something is, one must also understand how far away it is. If an object is close enough, astronomers can use parallax, the visual shift of an object as the observer's point of view changes. The closest star is our most obvious choice. The clever astronomer/physicist Friedrich Bessel first measured the distance to the Alphaa Cenaturi system in 1838. Bessel carefully observed the stars (which appeared as one in his instrument), and then waited for Earth to move in its orbit. Six months later, he charted the star again. Using the amount of its apparent shift against the more distant background stars, and knowing how far Earth had moved from one side of its orbit to the other (150 million km), Bessel was able to geometrically measure the distance by setting up a triangle. Alpha Centauri's movement was equivalent to viewing a sewing needle 11 km distant by looking out of one eye and then the other.

Parallax works well for objects in our cosmic neighborhood, and has been used to measure stars as far away as a few 100 light-years. But farther afield, the visual shift in objects is too small to measure. We need a different yardstick, and we find it in the nature of starlight. Every star sends out a set of light waves called a spectrum (pl. spectra). Spectra can tell us a star's temperature, what it is made of, and even how quickly the star is moving. The spectra of stars that have already been measured by parallax can be compared to distant ones. When the spectrum matches, researchers then measure the brightness of the star to see how far away it is.

The universe has provided another yardstick by way of variable stars, suns that brighten and dim in a regular pattern. Most famous of these are the Cepheid variables. The pulsing brightness of Cepheids relates directly to how hot – and therefore how bright – they are.

Astronomer Henrietta Swan Leavitt observed variable stars in 1908 in the Large and Small Magellanic Clouds, two companion galaxies of the Milky Way. Leavitt established a relationship between the period of variability of the Cepheids she was studying and their absolute brightness. By 1912, she was able to estimate the distance to the two galaxies.

Cepheid variables pulsate through periods of as little as a day to more than 70 days. Since their period betrays their intrinsic brightness, their dimness reveals their distance. Cepheids have given astronomers the distances to nearby star groups as well as to galaxies such as Andromeda and M81. Exploding stars can also be used to discern distance. The death throes of novae and supernovae unfold with some degree of predictability, supplying yet another measuring technique.

Beyond that, observers use the light from entire star clusters and galaxies to judge great distances. They compare close-in objects that can be measured in other ways, extrapolating out to the farthest reaches of the visible universe.

PLANETARY NEBULAE

Scattered throughout the cosmos are those puzzling planetary nebulae. To early observers, they appeared as spheres, but some had irregular edges, not like planets at all. Some even looked like comets, much to the chagrin of the comet-hunters. These things were on the mind of French astronomer Charles Messier. Messier's paramount interest entailed the search for comets, and he was good at it. In fact, the French King Louis XV called Messier "the ferret of comets." By the end of his life, Messier had discovered thirteen comets, along with thirty nebulae. On August 28, 1758, Messier made his first important mistake. He thought he had found a new comet in the constellation of Taurus. His subsequent observations proved that the fuzzy object was not moving like a comet, but rather staying put in its spot in the heavens. The misty distraction became the first entry in Messier's catalog for comet-hunters – the object Messier 1 or M1, known today as the Crab Nebula. Messier compiled a sort of "comet hunters beware" list of objects that might distract comet hunters from their searches.

Messier's catalog of comet-like sky phenomena would swell to over 100 objects. Messier said that he compiled his great directory "so that astronomers would no more confuse these same nebulae with comets just beginning to appear. "In addition to the Crab Nebula, Messier's Catalog includes M31 (the Andromeda Galaxy), M51 (the Whirlpool Galaxy), M8 (the Lagoon Nebula) and M45 (the Pleiades). Astronomers continue to use Messier's catalog as an important tool for searching the northern skies even today. Messier continued to work until only a few years before his death at the venerable age of eighty-seven.

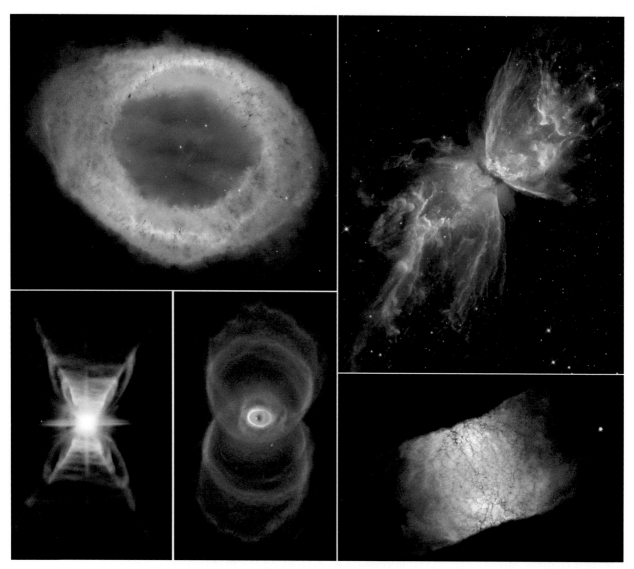

Fig. 6.4 An assortment of exotic planetary nebulae. Clockwise from top left: Ring Nebula (Images courtesy of NASA, ESA, and C. Robert O'Dell, Vanderbilt University), Butterfly Nebula (Image courtesy of NASA/STScI), the Retina Nebula (Image courtesy of NASA/ESA/STScI/ AURA), Hourglass Nebula (Image courtesy of Raghvendra Sahai and John Trauger, JPL, the WFPC2 team, and NASA/ESA), and the Red Rectangle (Image courtesy of NASA/STScI)

As time went on and astronomy matured, better instruments told the rest of the story of Messier's hazy sky smudges, a story of multi-colored expanding gases in complex patterns, glowing in baths of radiation. Many Messier objects are planetary nebulae. These roughly spherical clouds display wondrous and puzzling forms dictated by the way their host star generates them. Their names furnish an idea of their variety: the Eskimo, Butterfly, Cat's Eye, Helix, Dumbbell, Spirograph, Hourglass, Necklace, Ring, Egg and Lemon Slice.

One of the most picturesque examples of a cosmic gas bubble is NGC 7635. This is known popularly – and appropriately – as the Bubble Nebula. This gargantuan sphere is 10 light-years across. Near the cloud's center lies the star

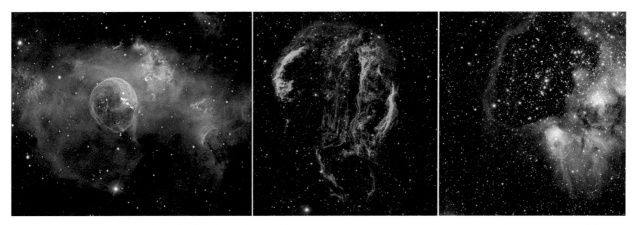

Fig. 6.5 Left: *The Bubble Nebula, pushed out by stellar wind from a massive star, spans 10 light-years across (Image courtesy of the University of Toronto).* Center: *The Cygnus Loop, a remnant of the titanic explosion of a supernova, drapes gas and dust across 130 light-years of space (Image courtesy of NASA).* Right: *An enormous superbubble seen in the nearby galaxy the Large Magellanic Cloud. A cluster of young, energetic stars floats at the heart of the formation (Image courtesy of ESO/Manu Mejias.)*

that made it, an O-type sun that shines with the light of several 100,000 Suns. Weighing in at 45 times the mass of our own Sun, the powerful star's savage winds thrust the bubble out through a surrounding cloud, causing it to glow.

The bubble in NGC 7635 is tame compared to the aftermath of a supernova. If a star is large enough, as in the case of a supergiant, it cannot hold together as it blows off its outer layers during its death throes. The explosion of gases is greater than that for a red or blue giant. Like the bow shock in front of a boat, a blast of radiation forces a shockwave to rush across the interstellar medium of drifting gas and dust. The Cygnus Loop is one such shockwave. This stellar tsunami traverses 130 light-years.

Nebulae provide us with clues of star death, but they also become nurseries for star birth. Stellar nurseries take on as many varied forms as do the planetary nebulae. We find the Eagle, Tarantula, Wizard, the Seven Sisters (the "Pleiades" or "Subaru") and Lagoon, to name a few. These elegant formations differ from planetary nebulae in that they are large clouds where stars are being formed, rather than the shells of dying stars.

The Eagle Nebula is a case in point. Its light-years-long pillars[4] of cold gas enshroud collapsing material in the process of condensing into stars. At the top of one of the columns (see Fig. 6.6), cool hydrogen gas combines with dust, becoming a star "incubator." Infant stars are embedded inside finger-like projections distending from the nebula's top ends. Each of the finger-like projections is larger than our entire Solar System. Knots of dark, dense gas – called EGGs ("evaporating gaseous globules") – likely contain stars in the early stages of formation, still swathed in the vapors of their creation. At the same time, stellar winds from nearby hot stars (off the image at upper right) are actually eroding the surfaces of the gas columns. Material can be seen streaming from the great clouds. The Eagle Nebula constitutes a race between star birth and radiation blowing away the very material that would enable suns to grow.

[4]The pillar at far left is roughly 4 light-years tall.

Fig. 6.6 This wide view of the Eagle Nebula shows the vastness of the interstellar cloud (Image courtesy of T. A. Rector, NRAO/AUI/NSF and NOAO/AURA/NSF; and B. A. Wolpa, NOAO/AURA/NSF), a closer view of the so-called "Pillars of Creation" in the central region (Image courtesy of NASA, ESA, and the Hubble Heritage Team [STScI/AURA]), and a detail showing fingers of material, each as far across as our Solar System (Image courtesy of Jeff Hester and Paul Scowen of Arizona State University, and NASA/ESA)

Another section of the nebula offers us a sense of just how large nebulae can be. The dark hydrogen cloud known as the Stellar Spire (Fig. 6.7) towers some 91 trillion km high. Cold hydrogen mixes with carbon and silicon dust, creating a long corridor of star formation. The spire is twice as long as the distance from the Sun to the Alpha Centauri system. Light from stars atop the column will not reach the base of it for nine and a half years. Like the rest of the Eagle Nebula, the spire is being torn apart by winds from hot stars off the upper edge of the image. The winds are both good news and bad news for infant stars. While they may erode some of the thinner gas/dust mix, keeping it from collapsing into a star, the blast of cosmic wind also bunches the thicker areas of hydrogen into dense waves where the seeds of stars may form (as seen at the top edge of the tower).

Stars form in groups within nebulae. Sometimes these groups remain together in families known as star clusters. When researchers find a group of massive, high-temperature stars, they know they have located a star cluster. Giant stars burn hot and die young, so they don't get the chance to drift far from the site of their birth. This means that smaller companions are still forming near them. Loosely organized "open clusters" often include thousands of stars and spread across 30 light-years of space. One of the most popular is the well-known Pleiades, or the Seven Sisters. This elegant blue cluster lies in the constellation of Taurus, and may be less than 100 million years old.

Globular clusters are another matter. These gigantic balls of stars contain some of the oldest suns in the entire universe. The vast spheres can hold more than a million stars. Up to 150 light-years across, their cores may contain 10,000 stars packed into a region just a few light-years in diameter. The stars swarm so closely together that they dance and swirl around each other in a gravity-ruled cotillion.

Using the main sequence as a yardstick, scientists have figured out ways to estimate the age of a star cluster. The Pleiades present a good case in point. A few hundred million years ago, the Pleiades would have been invisible from Earth. Drifting lethargically through the cold vacuum of space, it consisted of vast mists of dust and dark gas. The only hint of its existence would have been the few background stars that it blotted out. The great vapor was merely flotsam on a cosmic sea. But waves move across that sea. The great disk of stars that is the Milky Way triggers density waves that spiral out from the center, rippling through the dark clouds. Clots and kinks of material began to coalesce in the cloud as gravity waves moved through, triggered by passing stars or the shockwaves of nearby novae or supernovae. Eventually, one of those gassy eddies collapsed under its own weight, triggering nuclear fusion. Light flooded the gloomy wisps and whorls, and the Pleiades nebula flashed to life. The first stars sent out waves of energy that in turn triggered even more stars to form. Soon, the beautiful nebula took on the shimmering form we see today.

Among the thousands of stars of the "Seven Sisters," most flow along the main sequence path of the H-R diagram, discussed in an earlier chapter. But at the upper end, where hot, short-living stars should be, there are none.

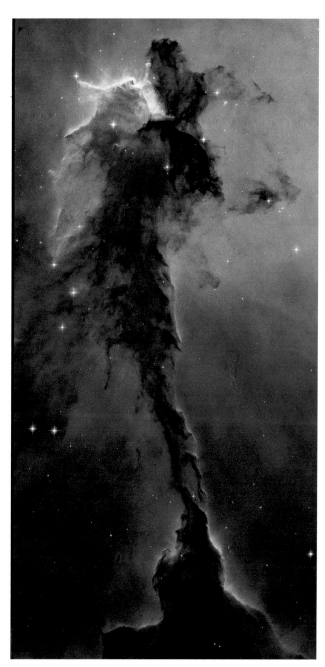

Fig. 6.7 *Long streams of stars begin their lives within the cold column of gases known as the Stellar Spire. This gargantuan structure, some 9.5 light-years high, is a small part of the Eagle Nebula (Image courtesy of STScI.)*

The stars of the Pleiades are just old enough that those stars are already burned out. Other stars along the main sequence farther down and to the left will burn longer because they are cooler. Eventually, some of these will become giants. The life span of B-type stars missing from the Pleiades is about 100 million years, so this gives us an upper age limit for this beautiful glowing nebula. Using the H-R sequence, we can date many star clusters. Figure 6.8 shows three examples, with the Pleiades in the middle. Judging from the place in the sequence where stars die off, we can date the Caldwell double-cluster in Perseus (squares) at 14 million years, the Pleiades at 100 million years (circles), and the NGC188, an open cluster in Cepheus, at about 7 billion years (triangles).

Fig. 6.8 The age of star
clusters can be determined by
their upper end-point in the
main sequence of the H-R
diagram. Star locations are
approximate (Art by the
author.)

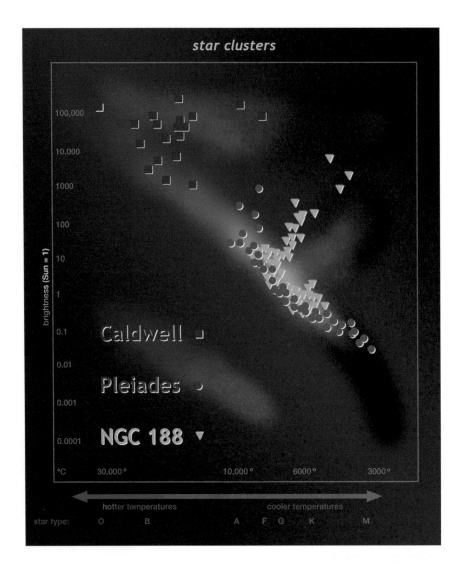

Fig. 6.8 The age of star clusters can be determined by their upper end-point in the main sequence of the H-R diagram. Star locations are approximate (Art by the author.)

A TRIP THROUGH THE CLOUDS OF SPACE

Not all nebulae are created equal. For a perspective on sizes of various nebulae, let's tackle a progression from small to large. We first zoom in on the inner rim of the Helix planetary nebula. Here, comet-like knots of gas stream away from the center of the nebula, each about as far across as our planetary system out to Neptune. The nebula may contain as many as 20,000 of these "knots." As we pull back, we see the Helix in all its incandescent glory. Just 700 light-years away, the Helix is one of the closest of the planetary nebulae. It stretches 2.5 light-years across.

We continue to move out to another part of the sky, where we see the next step in our progression, the Great Nebula in Orion. This nebula makes up one of the bright spots in the sword of the constellation Orion, the Hunter. Its flower-like petals stretch over 24 light-years. Unlike the Helix, which was born from a star's death throes, Orion is a stellar nursery where

Fig. 6.9 *Our planetary system seen out to Neptune, compared to the inner ring of the Helix Nebula (Image courtesy of NASA, ESA). The Orion Nebula (HST image) is ten times the size of the Helix, spreading its glorious clouds over 24 light-years. The Lagoon Nebula (ESO/VPHAS image) spans 110 light-years. Finally, the Tarantula Nebula stretches 1000 light-years from side to side (TRAPPIST/E. Jehin/European Southern Observatory image)*

many new stars are forming. At its heart lies the trapezium, a star cluster discovered by Galileo in 1617. Five of its brightest stars are huge, with masses between 15 and 30 times that of the Sun.

But the Great Orion Nebula is bested by the Lagoon, nearly ten times as large. Its vibrant tendrils snake through a region 110 light-years from border to border. Dark elliptical clouds – called Bok globules – dapple its billowing forms where stars are forming in collapsing gas. A colossal O-type star nearby forces some of the Lagoon's gas into a tornado-like formation.

Vast as it is, the lovely Lagoon is dwarfed by the Tarantula Nebula, a sprawling structure whose fluorescent gases pour across 1000 light-years of the cosmos. The nebula resides in the nearby the Large Magellanic Cloud (see the earlier section on galaxies in this chapter). A star cluster at its center causes the great cloud to glow. The stars in this cluster are so numerous that their mass is equivalent to 450,000 Suns. In 1987, the closest supernova ever observed with modern telescopes erupted in the suburbs of the Tarantula.

The hot giant stars forming within the nebulae today will explode within a few million years. Their death-blasts will send out new material for tomorrow's stars, along with shockwaves to compress the clouds around them, setting up the next generation of star formation. Around many of those stars, planets will form, some of which might host future generations of beings who will struggle to imagine the scale of the universe around them.

THE MILKY WAY, OUR OWN GALAXY

The nebulae around us stretch like strings of pearls along the arms of the Milky Way, a great spiraling disk of stars. Our Solar System, the Sun and all its planets, moons, asteroids and comets, drifts along a branch of stars between two arms of our galaxy's spiral, the Sagittarius and the Perseus arms (see Fig. 6.10). Specifically, we reside in a region called the Orion Spur, a trail of scattered stars and interstellar clouds. We have traveled along our local arm of the galaxy's spiral to visit many of the nebulae we have seen.

The Milky Way is a vast disk of 200–400 billion stars. It stretches some 120,000 light-years across. If the primary Solar System (out to the orbit of Neptune) were the size of a U. S. quarter (25-mm diameter), the Milky Way would be as far across as the continental United States. The Milky Way's central hub may be 7000 light-years thick, and glows with the reddish hue of dwarf stars, ghosts of vibrant suns born 10 billion years ago when the galaxy was a globe and not yet a flattened disk. A few bright young stars drift through the area, wanderers from the outer regions, ejected from the suburbs by close encounters with other stars.

Rivers of bright stars flow outward into a plain, mixing with dark clouds and lustrous nebulae. The disk thins out as it expands radially to some 50,000 light-years from the center. Newer suns in star-forming clouds dazzle the trails of material spiraling out from the inside. The stars do not circle around the galaxy as a racehorse on a track, but more as a carnival horse on a carousel. Stars bob up and down as they move within the rotating disk. The Sun will reach its roller-coaster peak 250 light-years above the galactic plane in 14 million years, at which time it will begin a stately star trek back down. Stars in our Sun's neighborhood orbit the galaxy once each 225–250 million years at a speed of about 230 km/s.

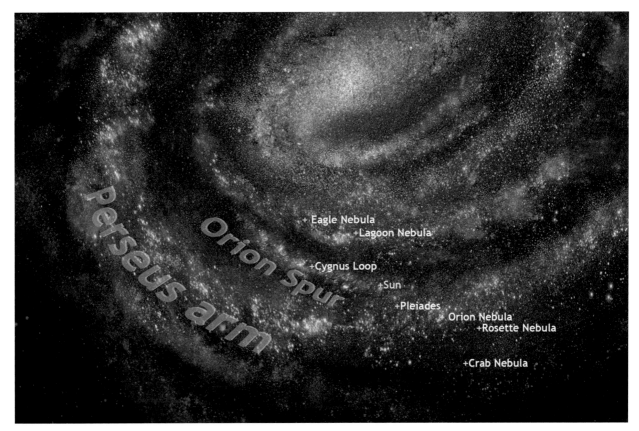

Fig. 6.10 Our own Milky Way Galaxy, with "nearby" objects mapped (Diagram by the author.)

The spiral pattern of a galaxy leads us to conclude that the arms of the Milky Way consist of trails of stars being carried along, as if on a giant pinwheel. But in fact, the arms we see are radiating waves like the ripples in a pond. Waves of newborn stars move from the inside out through the clouds of interstellar gas. The arms we see are simply the crests of the waves where gas is being packed together to form new suns.

Moving farther out, we enter a halo of globular clusters, spheres of stars. About 200 globular clusters orbit the galaxy. The Large and Small Magellanic Clouds, disorganized gatherings of stars, now pass by us. Like the globular clusters, they orbit the Milky Way in odd, rosette-shaped patterns. The Magellanic galaxies swoop in as close as 180,000 light-years from the galactic center. On the other side float Sculptor and Fornax, dwarf galaxies. Two smaller groupings, Leo I and Leo II, round out our local group.

The Andromeda Galaxy, glowing majestically at a distance of 2.5 million light-years, was long thought to be a twin of the Milky Way. Its elegant star arms spiral out from a central hub of ancient stars, and observers could see, in between our interstellar clouds, a similar hub at the center of our own galaxy. A bright bar crosses the center of some spiral galaxies, affording them the moniker "barred spirals." It turns out that our own galaxy is barred, rather than a simple spiral like Andromeda.

The interior of our star island is riddled with incandescent clouds, stars small and giant, old and new, and mysteries at its core. Let's venture out to see these wonders. Or, as Jules Verne might have put it, let's journey to the center of the galaxy.

From our suburban neighborhood, the center of our Milky Way Galaxy lies some 28,000 light-years distant. Such a trip seems unlikely at best, as our farthest human journey – to the Moon – took 3 days to transverse a mere quarter of a million miles. The fastest-moving artificial object to date,[5] NASA/JHUAPL's New Horizons Pluto spacecraft, made the 4.76 billion-km trip to Pluto in 9 years and 5 months. At its top speed, our journey to the center of the galaxy would take a time investment of 449 million years.

The trip may sound impossible, but when a traveler approaches the speed of light, or even a significant fraction of it, time seems to slow down. An astronaut traveling at high speed might experience a time period of a few years, while back on Earth a few millennia have passed. And what a voyage it would be! As our ship begins its voyage, we pass Mars, the asteroids, and the gas and ice giants quickly. The Kuiper Belt extends farther, and the Oort Cloud is with us for even longer, despite our increasing speed. Once out of the Oort Cloud, things seem to slow down. Our ship continues to accelerate, but the nearest star is so far away that we do not pass it for some years. We gain speed, and the stars begin to crawl slowly by. Familiar constellations drift into new patterns. Behind us, the Sun and all that we have ever known become a faint dot among many stars, fading into the warping constellation Taurus.

Our journey follows the plane of the galaxy, the equator of the Milky Way's great disk. We pass nebulae and newborn stars, and then exit the Orion Spur of the Perseus Arm. As we pass between arms of the spiral galaxy, the sky becomes dark with clouds and a lack of stars. But as we enter the next arm in, the Sagittarius, we again see a chain of colorful nebulae, stellar nurseries. We sail between the Lagoon and Eagle nebulae, glowing with the fires of infant suns. We watch the dance of multiple-star systems, the death of supergiants, the cold glow of white dwarf stars and the ember afterglow of red dwarfs as old as the galaxy itself. Finally, we enter the central hub, an enormous dome of ancient, golden stars. Here, we get a sense of the form of the galaxy, not a perfect spiral but rather a spiral with a kink in each of the two great arms emerging from the central swirl of stars.

Gravity urges us on toward the mysterious core, obscured from Earth by pathways of dark interstellar dust. Gargantuan glowing tendrils trace out fierce magnetic field lines that snake throughout the core region, threading through the many reddish-orange elderly stars here. At the core simmers a powerful radio source called Sagittarius A* (pronounced "Sagittarius A star"). It is unlike any other radio source in the entire galaxy. Sag A* has a mass of roughly four million solar masses packed into a region slightly larger than our Solar System. Around it swarm several hundred stars, crammed into a sphere just a few light years across. The racing stars in such tight orbits suggest that a massive black hole lies at the center of our Milky Way.

[5] Launch velocity was 16.26 km/s. A Jupiter gravity assist added 4 km/s. The *Voyager 1* spacecraft is actually now traveling slightly faster due to gravity assists from both Jupiter and Saturn, but New Horizons got there on its own power.

If a black hole does, in fact, sit enthroned at the center of the great Milky Way Galaxy, it does not behave as other black holes do. Most appear to be surrounded by whirlpool-like accretion disks, swirling patterns of material emanating prodigious amounts of X-rays. This is not the case with the misbehaving Sag A*, whose X-ray emissions are fairly faint and calm. The mystery of Sag A*, at the heart of the Milky Way, awaits further data and advances in our technology.

BEYOND THE MILKY WAY?

At the start of the twentieth century, observers could not distinguish between nebulae – those beautiful stellar nurseries or marks of exploded stars – and galaxies, great collections of millions of stars. Some of the neb-ulae seemed to have a consistent structure, and these were called "spiral nebulae." At the time, astronomers engaged in heated debates about the scale of the Milky Way itself. Was it the main part of the universe as a whole, with spiral nebulae simply another type of gas cloud like the Pleiades or Horsehead? Or were these spirals island universes like the Milky Way, but far enough away that their stars blended into a soft shape?

The debate was solved, once and for all, at the eyepiece of the 100-in. telescope of Mt. Wilson Observatory northeast of Los Angeles, California. At the helm was Edwin Hubble, who took a series of photographic plates of the "spiral nebula" M31. His images clearly showed that the spiral shape was made up of thousands of stars. He also spotted several Cepheid vari-able stars, later used to establish the actual distance of M31, now known as the Andromeda Galaxy. Hubble calculated that the distance to Andromeda was far greater than the size of the Milky Way, proving that Andromeda was an entirely separate galaxy akin to our own.

GALACTIC MENAGERIE

Eight out of every ten galaxies holds the shape of a flattened disk. The majority of these are called spiral galaxies, because their stars take on a swirling, spiral pattern. Some spirals are tightly bound, while others have loosely wound arms and small, faint nuclei. Like the Milky Way, NGC1300, pictured in Fig. 6.11, is a barred spiral. It lies 69 million light-years away, and is 110,000 light-years across, making it about two-thirds the size of our own galaxy. At the center of NGC 1300s bar is a spiral structure 3300 light-years in diameter. Astronomers call this spiral-within-a-spiral for-mation a "grand design" structure. Grand design structures are only found in barred spirals with a large-scale bar.

The elegant Pinwheel Galaxy, also known as M101, is a classic spiral, seen face-on from Earth. Just 21 million light-years away, this magnificent helm of stars stretches 170,000 light-years across, making it about the size of

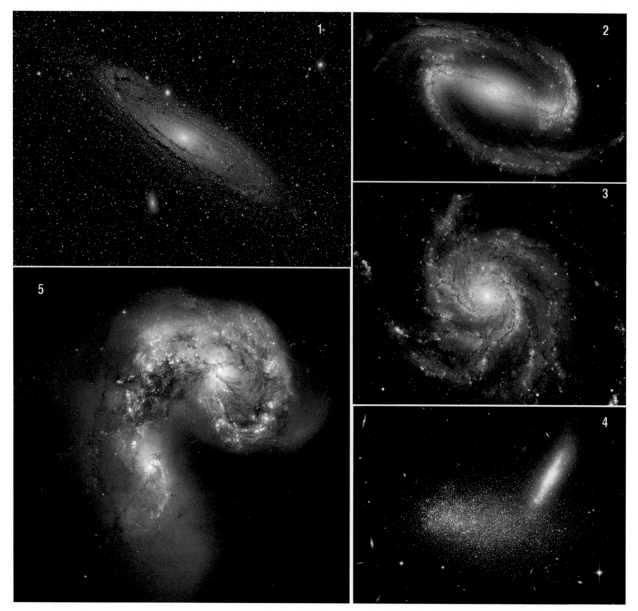

Fig. 6.11 Clockwise from upper left: **1.** *The Andromeda Galaxy (Image courtesy of Adam Evans, Wikipedia Commons; https://en. wikipedia.org/wiki/Andromeda_Galaxy#/media/File:Andromeda_Galaxy_(with_h-alpha).jpg).* **2.** *NGC 1300, a barred spiral galaxy (Image courtesy of Space Telescope Science Institute.)* **3.** *The Pinwheel Galaxy (Image courtesy of European Space Agency & NASA). (Acknowledgements: Project Investigators for the original Hubble data: K. D. Kuntz GSFC, F. Bresolin, University of Hawaii, J. Trauger, JPL, J. Mould, NOAO, Y.-H. Chu, University of Illinois, Urbana, Davide De Martin, ESA/Hubble.)* **4.** *The irregular galaxy PGC 16389 partially covers a more distant spiral galaxy (Image courtesy of ESA/Hubble & NASA, Luca Limatola).* **5.** *The Antenna Galaxy (HST image courtesy of the Space Telescope Science Institute.)*

our own galaxy. The Chandra X-ray observatory has located a powerful X-ray source at its center. Further studies have revealed that the core object is a binary star with one component being a black hole with 20–30 solar masses.

One of the most remarkable central bulges rears up from the disk of the Sombrero Galaxy. Seen almost edge-on, this dusty behemoth is surrounded by a soft halo. Dark gas and dust ring its disk, dramatically defining the plane of the galaxy.

Some spiral galaxies have bright, point-like cores emitting continuous streams of deadly radiation. Seyfert cores are as intensely bright as the entire Milky Way, compressed into a small nucleus. The Whirlpool Galaxy, M51, is an example of a Seyfert galaxy. The Whirlpool's pole points directly at us, displaying the full glory of M51's spiraling, starry arms. With a diameter of 86,000 light-years, this beautiful spiral is about a third the size of the Milky Way. At its heart resides a black hole surrounded by two interweaved rings of dark dust.

Spiral and barred spiral galaxies are only two of many forms that galaxies take. Elliptical galaxies get their name from their nearly featureless oblong shape. While the stars in a spiral galaxy circle majestically around a hub, those in an elliptical galaxy move in random directions around the center. The 'dwarf spheroidal ellipticals' span diameters of less than 325 light-years across, while 'giant ellipticals' are the largest known galaxies, some stretching beyond 300,000 light-years, with 100 times the mass of the entire Milky Way. Ellipticals may be more ancient than most spirals, on average, as they seem to be starved of the gases that lead to star formation.

One elliptical galaxy tallies among the largest structures in the universe. Two hundred times as massive as the Milky Way, supergiant M87's stars reach a diameter of 980,000 light-years, traveling at great velocities near its center. Stars in its core follow compact, rapid orbits around an unseen central object with 5 billion solar masses. Unseen, that is, in visible light. But lurking at M87's heart, a supermassive black hole spews out a 5000 light-year-long stream of material at nearly the speed of light. This jet is wobbling, so that its material forms a spiraling helix. Within this superheated radioactive beam, blobs of material fan out some 250,000 light-years. Each day, material equal to 91 Earths falls into the black hole. Thin filaments 100,000 light-years long snake through the galaxy. An eruption 70 million years ago formed a vast empty bubble in M87's hot interstellar gas, adding to the remarkable portrait of the enormous galaxy.

Falling in size between ellipticals and spirals are the lenticular, or lens-shaped galaxies. Lenticular galaxies may be an intermediate form between spirals and ellipticals because they contain more interstellar gas than ellipticals, but less than spirals. Lenticular galaxies resemble spiral galaxies stripped of their arms, possessing only the nucleus and halo.

Even more disorganized are the "irregular" galaxies. Some of these might be scrambled spirals, retaining just hints of an earlier structure. Others exhibit bizarre appearances that don't fit any category. Most resemble the disks of spiral galaxies, white and dusty. The colors of irregulars show that most stars are young and massive, like blue giants or supergiants. The farther from us we observe, the more irregular galaxies we see. Because light from distant regions is older, this shows us that irregular galaxies were more common in the early days of the universe.

Fig. 6.12 The unusual galaxy
called "Hoag's Object"
resembles a planetary nebula
but is in fact a vast ring of
stars surrounding a galactic
nucleus of redder, older stars
(Image courtesy of STScI/
AUR.)

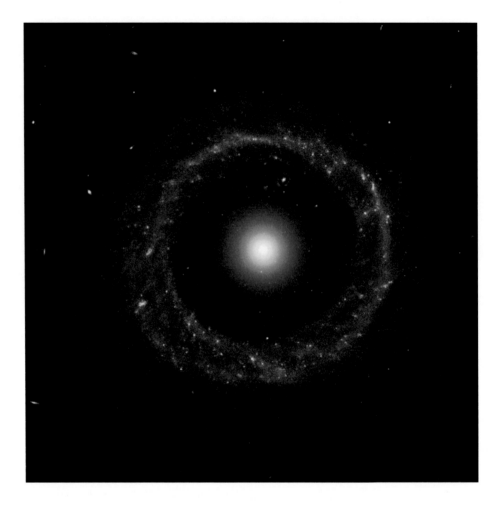

Another peculiar galaxy, discovered by astronomer Art Hoag in 1950, is now known as Hoag's Object. Spanning some 120,000 light-years across – slightly larger than the Milky Way – the entire galaxy resembles a ringed planet. The galactic nucleus is yellow, warmed by the light of ancient red stars. A blue ring of young star clusters encircles the nucleus, with a remarkable gap separating it from the central hub. Within the ring are giant young stars.

The collision of galaxies often results in beautiful cosmic scenes. One example is the Antennae Galaxy, named after its two long, curving trails of stars extending beyond the two colliding galaxies. (These long streamers give the appearance of an insect's antennae.) Originally, the galaxy was two separate galaxies, a spiral and a barred spiral. Roughly 600 million years ago, the two great islands of stars began to pass through each other. Computer models estimate that the two galaxies will completely merge, with the result being an elliptical galaxy.

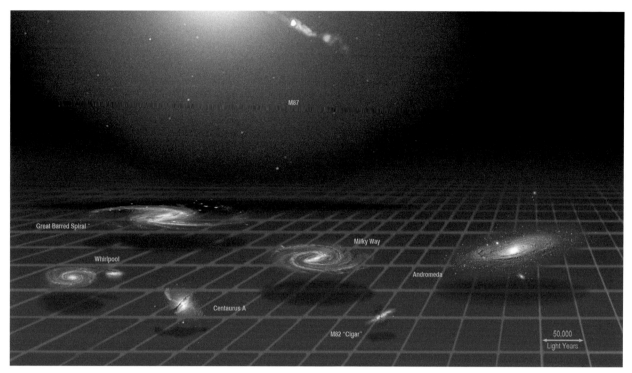

Fig. 6.13 A selection of galaxies compared. Each square on the grid is 50,000 light-years across. The smallest galaxy pictured is the irregular M82, or "The Cigar." The largest is the behemoth M87, spanning 980,000 light-years across. Its relativistic jet emanates from its center, just off the page at top, toward the lower right, issuing from the massive black hole at its core (Some galaxies modified from Hubble Space Telescope images, courtesy STScI; art by the author.)

GALACTIC CLUSTERS

We have stepped from Solar System to nebula, from nebula to star cluster, and from star cluster to spiral and elliptical galaxy. Our next level of scale is the galactic cluster. Like stars, galaxies often form in groups, ranging from a few loosely held together members to great clusters containing thousands of galaxies in many forms. The arrangement of these clusters gives us insight into the structure of the entire universe. Galaxies are arranged as if they ride the surfaces of bubbles in a pile of soapsuds (see Fig. 6.14). Between the complex web of galaxies lies great empty cavities of space.

The Milky Way is part of the Local Group, a small family of 54 galaxies, most of them dwarfs. The Local Group is dumbbell-shaped, traversing ten million light years of cosmos. The Local Group, in turn, is part of the Laniakea Local Supercluster.[6] Laniakea includes 100,000 galaxies 520 million light-years side to side.

At the center of the Laniakea Supercluster lies a great mystery. Across hundreds of millions of light-years, an unseen force affects the movement of galaxies. Called the Great Attractor, it is a concentration of gravity tens of thousands of times that of the entire Milky Way. The Milky Way Galaxy is racing toward this region at a breakneck speed of 600 km/s, perhaps under the influence of the Great Attractor's gravity.

[6] The Milky Way's Local Group was once considered part of the Virgo Supercluster, but Virgo is now counted as an offshoot of the Laniakea Local Supercluster.

Fig. 6.14 *The frothy structure of the universe is seen in this computer-generated map based on findings of the COBE and WMAP satellites. Each dot is an entire galaxy, and those galaxies string together like the bubbles in soapsuds (Image courtesy of NASA.)*

The Virgo Cluster, the arm in which the Milky Way resides, contains between 1300 and 2000 spiral and elliptical galaxies, including M87. Other nearby clusters include the Fornax, Antlia, Centaurus and Hydra clusters.

Galactic superclusters do not close out our progression of cosmic scale. The grandest structure yet seen in the universe is 7.2 billion light-years across and 10 billion light-years long, with a depth of nearly a billion light-years. This mind-boggling formation is known as the Hercules/Corona Borealis Great Wall. The Great Wall appears to contain vast masses of material. Much of it may be dark matter, a theorized form of matter that astrophysicists deduce from its effects on visible matter, radiation and large-scale structures of the universe. The general soap bubble pattern of the universe may, in fact, be due to streams of dark matter, clumping galaxies along otherwise invisible pathways.

QUASARS

At the farthest fringes of space and time float mysterious, powerful, star-like objects. These energetic sources are incredibly bright, and their red shifts betray great age and distance. Prodigious amounts of radiation pour from them, and many eject jets of material from deep within their cores. Many exist at a range of nearly 13 billion light-years, which means that they formed within less than a billion years of the universe's genesis. Some pump out as much light as a thousand Milky Way galaxies, but their light comes from a focused source, more like a star than a galaxy. Hence, they are referred to as "quasi-stellar objects," or quasars.

Many researchers were initially skeptical that quasars could be so distant and so bright, but advancements in instruments have proven the doubters wrong. Quasars lurk at the centers of incredibly distant and ancient galaxies. The majority inhabit a region of space halfway to the edge of the visible universe. The lines in their light spectra are moved down the "rainbow" more than three times what they are in a normal spectrum from an object standing still. This means that the light from quasars raged forth when the universe was less than a third as old as it is now.

Quasars are probably compact regions in the centers of massive galaxies whose cores enshroud a supermassive black hole. The accretion disk of the central black hole powers the quasar's prodigious energy. Often, two titanic beams of material burst out of the galaxy's nucleus in opposite directions, with their material moving at nearly the speed of light. They typically race through space well beyond the edge of the galaxy itself, sometimes extending as far as a million light-years from the core, where they ram into intergalactic gas. As the jets enter intergalactic gas, they fan out like a waterfall hitting a pond.

Related to the quasars are blazars. Like many quasars, they generate jets of material moving at relativistic speeds (near the speed of light), but in the case of blazars, their polar jets are pointed directly at Earth. Often, these jets are surrounded by superluminal features, the fingerprints of shockwaves moving out from the polar jets. While the jets are traveling close to the speed of light, the particles within them are also emitting light. Their light approaches the observer at nearly the same rate as the jet itself. The light emitted over a matter of centuries at the front end of the jet reaches us at nearly the same time as that from the base. This sets up an optical illusion. It appears that light from the more distant sections is moving more rapidly than that from the front, imparting the impression that those particles are traveling faster than the speed of light.

The galaxy Centaurus A is much like a quasar, but it is close by. By making radio observations of the polar jet a decade apart, astronomers have determined that the inner parts of the jet are moving at about one half of the speed of light. The radio jets of Centaurus A are over a million light-years long. As they slam into intergalactic gases surrounding the galaxy, they generate fierce X-rays. Within a bizarre warped disk of dark material crossing its center, observers have spotted recent star formation. In fact, they have located over 100 stellar nursery regions in the disk.

The distant quasars may resemble Centaurus A up close, but their vast distances challenge our very best instruments. We would love to be able to study galaxies even farther away, star islands at the edge of space and time. But how can we? Actually, despite the limitations of our own technology, there is a way…

Fig. 6.15 The galaxy Centaurus A in visible light, above, and in radio wavelengths seen via the Very Large Array telescope, a group of radio dishes in New Mexico. Note the strong jets of material seen clearly in the radiation they give off, viewed below (Images courtesy of NASA/VLA.)

A WEIGHTY MATTER

Albert Einstein's mathematics predicted that heavy objects actually warp the fabric of space and time (two aspects of our universe that are interrelated in quantum physics). The theory of relativity declares that if an object is heavy enough, it will warp space around it in such a way as to bend the light from an object beyond it. In fact, observers confirmed this phenomenon in 1919, during a total eclipse of the Sun. As the Moon blocked the Sun's light, stars became visible. Observers the world over carefully documented the position of the stars near the Sun during the eclipse. The accurate measurements showed that stars seemed to disappear behind the Sun later than they should, and appeared from behind the Sun earlier, showing that their light waves had been bent by the Sun's gravity. In other words, the Sun acted like a cosmic magnifying glass. This warping phenomenon is called gravitational lensing.

Theoretically, it should be possible to see a distant object hiding behind a black hole or closer galaxy. Its light will be pulled around the heavy foreground object to form an image beside it. In a famous 1936 paper, Einstein proposed just that, but went on to comment that, "Of course, there is no hope of observing this phenomenon directly." Einstein was seldom mistaken, but in this he was wrong.

One of the most dramatic sources of gravitational lensing, imaged by the Hubble Space Telescope, is the galaxy cluster Abell 370. In Fig. 6.17, the bright globe-like shapes are Abell's galaxies. The bizarre arc to their right is actually a sort of mirage, consisting of two warped images of a galaxy beyond the Abell cluster. The streaks arching around a point at about the center of the image are distant galaxies whose light is warped around the immense gravity of Abell's galaxy group.

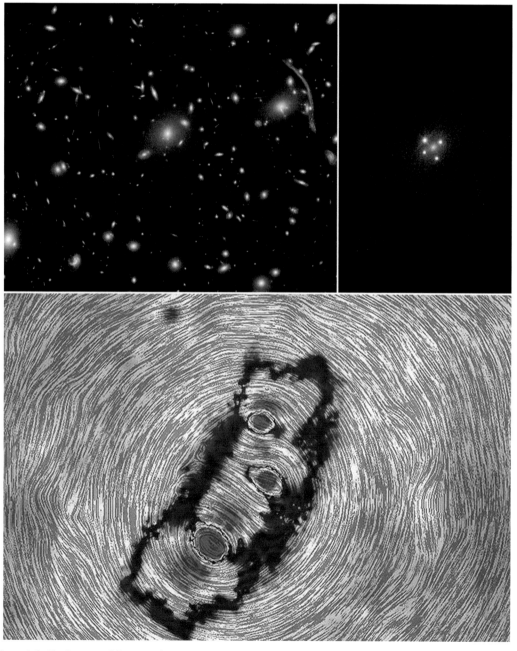

Fig. 6.16 Above left: *The images of distant galaxies warp around the invisible gravitational lens Abell 370.* Upper right: *The Einstein Cross.* Bottom: *An HST data map of the Abell 2744 galaxy cluster's gravitational warp of surrounding space (HST images courtesy of STScI; map image courtesy of Marusa Bradac, UC Davis physics group.)*

Fig. 6.17 The expansion of the
universe can be likened to the
expanding dough in baking
bread, with the raisins
representing galaxies (Art by
the author.)

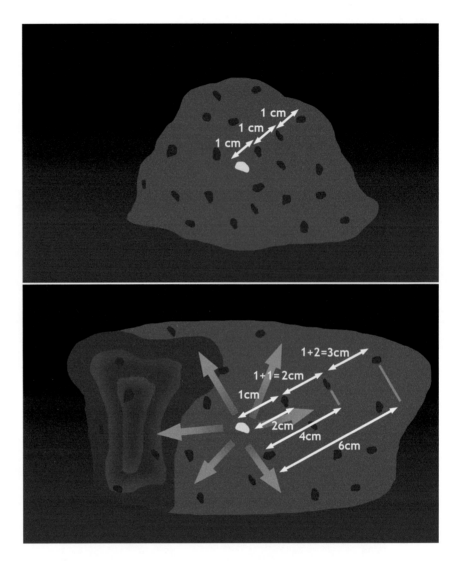

A classic example of gravitational lensing is known as the Einstein
Cross. A faint foreground galaxy, nearly invisible in the center, warps space
around it in such a way that the more distant quasar's light bends into four
separate images, creating a ghostly cloverleaf.

Advances have enabled researchers to map the complex interplay of
these gravitational lenses. Another lens in the Abell 2744 galaxy cluster
contains the mass of 2 quadrillion Suns. Researchers using Hubble Space
Telescope data charted the effects of Abell 2744s massive warping of space
and time around it. Spanning three million light-years across, their map
assigns red to areas most strongly warped, amplifying light some 30 times.
Gray lines show the direction and angle at which images of distant galaxies
get warped. Abell 2744 has provided astronomers with a glimpse of the
very ancient universe – a triple echo of a galaxy so distant that it existed
just 500 million years after the birth of the universe.

Gravitational lensing opens a window into the very distant universe,
a realm so far away that we might not be able to detect it without this natural magnifying glass given to us by immensely massive objects nearby.

Chapter 7
Understanding Scale in the Universe

Educators understand how difficult it is for the human mind to comprehend the vast numbers associated with astronomy. They've been using basketballs, marbles and pins to create models of the Solar System for decades. One popular model is known as the "Pinhead Solar System." Because of the difference between the scale of planets and the scale of their locations around the Sun, this particular model is displayed in two stages. In the planetary stage, the Sun is the reference point, represented by a basketball. At this scale, Mercury is represented by the diameter of the shaft of a straight pin. Both Venus and Earth show as the head of the pin. Mars is the equivalent of a granule from a common cold capsule or antihistamine. Jupiter and Saturn can be seen as large marbles or beads. Uranus and Neptune are the size of ball bearings or an M&M candy or equivalent. Pluto, on this scale, is the size of the period at the end of this sentence.

Distances are another matter. With the Sun as a basketball, Mercury will be 11 m/36 ft.[1] away, so most presenters opt to change the scale, using string or colored rope to exhibit distances on a long line away from the Sun. If we shrink the Sun to the size of the head of a pin, Mercury is now microscopic (as are all the other planets) at a distance of 10.6 cm/4.2 in. along our string. Venus clocks in at 19.8 cm/7.8 in., with Earth and Mars at 27.4 and 41.6 cm/10.8 and 16.4 in., respectively. From here, Johann Bode's gaps become obvious. Jupiter orbits the Sun at 1.42 m/4 ft., 8.2 in. away, leaving a dramatic gap where the main Asteroid Belt rings the system. Saturn lies at 2.57 m/8 ft., 4.9 in. Uranus revolves out at 5.26 m/17 ft., 3 in., where most of the air common to our experience will freeze into a solid block or puddle on the surface as a liquid. Neptune's orbit shows up at a dramatic 8.2 m/27 ft out, and the farthest extent of Pluto's is 10.83 m/35 ft., 6.4 in. It is important to keep in mind that each of these tick marks are spots on a circle,[2] and that circle extends as far to the other side of the Sun as the visual model exhibits on this side.

Basketballs and colored string constitute a great way to communicate in the classroom, but more sophisticated Solar System models have been constructed throughout the world, many in the outdoors. The Colorado Scale Model Solar System is one such model. This assemblage shows both planetary sizes and distances between them using a 1–10 billion scale. Located in the shadow of the Rocky Mountains in Boulder, Colorado, the model begins at the Fiske Planetarium on the campus of the University of Colorado, and moves out into the community along streets and pathways. Referred to as a "walkable scale model," it utilizes true scale. In other words, the model uses the same scale for the distances in the Solar System and for the planets themselves. The model is the brainchild of astrophysicist, educator and author Jeffrey Bennett. Bennett's inspiration began with his teaching of elementary school children. "NASA puts out these photo montages, and I would show them these and they thought this was how space really looked. I realized that if I was going to give them any kind of real

[1] For educators, these units appear in both English and metric forms.

[2] The planets actually follow elliptical orbits, but for this scale a circle is close enough.

understanding, they needed to have those beautiful pictures accompanied by a sense of scale."

Bennett started doing model Solar Systems with his school kids, using clay balls for the inner planets and paper maché over balloons for the outer planets, but his students had trouble mentally combining the planets' sizes with the distance scale. "I started to see what kind of scale we could come up with to see both. One to ten billion came out to be a great compromise. It's less than a half mile to Pluto on that scale." Bennett began teaching at the college level, and realized that the college students were under the same misconceptions as the younger students. In 1984, Bennett and a group of students proposed a scale model for the University of Colorado. Several times, their presentation got postponed. "We were on the schedule to present in February of 1986. That January, Challenger blew up." CU's commissioners decided that the project would be a perfect memorial for astronaut Ellison Onizuka, a CU alumni, and the other Challenger astronauts. The student project became a $30,000 sophisticated memorial.

In Bennett's model, the Sun is 14 cm in diameter (the size of an average grapefruit). Planets range in size from specks of dust to marble-sized Jupiter, each a bronze display mounted on a granite pedestal. At this scale, Earth is the size of a small pinhead. The terrestrial planets (Mercury, Venus, Earth and Mars) can be found on pedestals within a few dozen steps of the Sun, with Earth standing 15 m away. The gas and ice giants spread out across a grassy park and into the adjacent town. A walk from the Sun to Pluto is 600 m, or a third of a mile, about a 10-min hike without stops. Bennett says, "You can fit Earth and the entire orbit of the Moon in the palm of your hand – which represents the farthest humans have ever traveled." At the scale of Bennett's model, the Alpha Centauri star system – the closest stars to Earth – are 4000 km away, roughly from Boulder, Colorado, to the Panama Canal.

Fig. 7.1 Jeff Bennett's scale model of the Solar System begins in front of Fiske Planetarium (left) and wanders across the Boulder cityscape (Photo/diagram courtesy of Jeff Bennett.)

The model on the CU campus was one of the first in the United States, providing inspiration for a series of models called Voyage. The Voyage models combine science with modern sculpture. Each planet is encased in plexiglass, along with color photos. The first of the Voyage replicas debuted at the Smithsonian Institute in October of 2001. This model spreads out along the National Mall in Washington, D.C. The Smithsonian's model is only the first of a series, with six more either currently planned or in place throughout the country.[3]

The city of Gainesville, Florida, offers a more lyrical approach to the scale Solar System model. Stretching along nearly a mile of uninterrupted roadway, its planetary monuments are constructed from recycled materials. Gainesville's model, the brainchild of local artist Elizabeth Indianos, has the advantage of visibility. It can be viewed from a single vantage point. No business signs or driveways intrude upon it. An information plaque emblazons each planet's pedestal, showing relative size compared to the Sun, distance from the Sun, and other information. The scale is 4 billion to 1.

The world's largest scale model of the Sun's family stretches across 300 km of Swedish countryside. The model starts with the Ericsson Globe football arena, in the heart of Stockholm. At a diameter of 110 m, it is the largest hemispheric building in the world, and serves as the Sun on this scale. The inner planets also reside in Sweden's capitol, with a 65-cm Earth standing in the Swedish Museum of Natural History at a distance of 7600 m. A full 40 km from the arena, Jupiter and some of its moons are represented in a flower garden. Jupiter is 7.3 m across. Neptune, last of the giants, inhabits the fishing village of Söderhamn, some 229 km from the globe. Its location gives a nod to the deity Neptune, god of the sea. Neptune's acrylic globe is illuminated in blue light at night. The Swedish model also includes some of the objects near the Kuiper Belt, such as the ice dwarf planets Pluto and Eris, as well as the smaller bodies Ixion, Saltis, and distant Sedna. Halley's Comet is also given a place. The termination shock, the edge of the Sun's influence in interstellar space, is even marked out at the Institute of Space Physics in the town of Kiruna, some 950 km away.

[3] For more information, or to set up a Voyage model in your community, see http://voyagesolarsystem.org/program/program_default.html.

Fig. 7.2 Left: *Stockholm's Ericsson Globe Arena stands in for the Sun in Sweden's Solar System model (Photo courtesy of Fredrik Posse/ Stryngford Photo/via Wikimedia; https://commons.wikimedia.org/wiki/File:Globen_Stockholm_February_2007.jpg)* Center: *The Neptune model in Söderhamn (Photo by Vincenetas via Wikipedia Commons; https://commons.wikimedia.org/wiki/File:Neptune_ model_of_Sweden_solar_system.jpg)* Right: *The Sedna model, erected in the city of Luleå in northern Sweden (Photo Dag Lindgren via Wikipedia Creative Commons; https://commons.wikimedia.org/wiki/File:Sweden_Solar_System_-_Sedna_2.JPG.)*

Fig. 7.3 *The Galaxy Garden helps visitors to understand our place in the Milky Way. Clockwise from left: Aerial view of the garden; diagram showing nebulae, each represented by a specific flower; a fountain represents the black hole at the galaxy's center. Note its form, which echoes the polar jet (water), gravity well (pedestal), accretion disk and event horizon (the swirling pool at its base) (Images courtesy of Jon Lomberg, aerial image by Pierre and Heidy Lesage. Used with permission.)*

Roughly 50 permanent models of the Solar System have been erected across the United States, Canada, Australia and Europe, with the United Kingdom and Germany each hosting four.

Perhaps the draw of these models is their other-worldly essence, which enables us to understand the corners of a universe far grander than our daily experience. But for some, the Solar System is not enough. Space artist/writer Jon Lomberg has designed a 100-ft. diameter garden in the shape of the Milky Way Galaxy. This outdoor scale model displays various landmarks of our galaxy in the form of indigenous Hawaiian plants. Tropical flowers represent nebulae. At the garden center, a fountain symbolizes the black hole at the hub of the Milky Way. The garden is on the site of the 9-acre Paleaku Peace Gardens Sanctuary in Kona, Hawaii. In the Galaxy Garden, one foot equals a thousand light-years. The Sun would be the size of a virus, and the entire Solar System out to the far edges of the Oort Cloud would extend only one hundredth of an inch across. Says Lomberg, "All the stars you can see with the naked eye are on the same leaf as the Sun, or on neighboring leaves. The awe-inspiring countless stars on a dark night are merely our closest neighbors, sharing a small branch of the much larger Galaxy Garden."

Lomberg's spectacular scale model of the Milky Way calls to mind the thoughts of William Herschel, discoverer of the planet Uranus. Herschel drew a comparison between the diverse plants of a garden and the wondrous formations of the universe around us. The dedicated astronomer likened the star-birthing nebulae, and the stars in contrasting stages of life, to "a luxuriant garden, which contains the greatest variety of productions; in different flourishing beds…[I]s it not almost the same thing, whether we live successively to witness the germination, blooming, foliage, fecundity, fading, withering, and conception of a plant, or whether a vast number of specimens, selected from every change through which the plant passes in the course of its existence be brought at once to our view?"

Part II

Our Place in Time: How Our Concepts
Have Evolved

Chapter 8
The Worlds Around Us

Humans exist and operate in four dimensions: height, width, depth and time. Up to this point, we have been exploring our universe primarily within the first three dimensions, where we can move freely. But the fourth dimension, that of time, is, for us, strictly a one-way affair. When we are born, we are thrust into the river of time, and its currents pull us – and our history – inexorably along into the future. Although the work of both Albert Einstein and, more specifically, Steven Hawking, implies that travel through time is possible, Earth's inhabitants currently have no means to go in any direction but forward. But we have seen how huge objects with strong gravity fields actually warp not only physical space but the fourth dimension as well.

While our past chapters have explored the physical scale of our universe, we now turn to that fourth dimension. In the scale of time, we see how our view of the cosmos has changed and matured. Knowledge has informed our concept of planets, moons and stars. But beyond that, we can also witness the evolution of our conception of how we might travel within the great scales of our physical universe. And just as our trek through moons, planets, stars and galaxies revealed much about the place in which we live, this trip through the scale of time is a voyage of discovery.

As our gaze has reached farther into the glowing splendor of the universe, we have also traveled back in time. While we resolve light emanating from ever-more remote objects, we see light that is increasingly ancient (see Fig. P.6). It is only fitting, then, that we travel back in time just a short ways to compare our modern view of the cosmos with the visions of earlier philosophers, scientists and artists.

To the ancients, the stars seemed to stay in their proper places, affixed to the night sky, working their way in a northerly or southerly march as the seasons came and went. All was organized and well behaved, with the exception of a few nomadic stars. These they called the "wanderers," or planets.

As time and knowledge progressed, the naturalist/philosophers became bona fide scientists, and as their view of the skies matured, they came to realize that those wandering stars were other worlds. Primitive telescopes resolved them into disks, and soon the observers understood those disks to be spheres, real worlds, perhaps not unlike Earth.

Preeminent in the night sky, Earth's Moon is close enough for even a casual observer to glimpse its grosser features without the aid of a telescope. To early observers, it was obvious that the Moon's dark regions were surrounded by higher ground, leading to the logical analogy of Earth's dark oceans and bright continents. The idea was not far off in concept, but a bit off in time. The dark lunar "maria" (oceans), bays and seas were, at one time, oceans, but they were not filled with water. These primordial seas brimmed with molten rock. Vast magma seas cooled and dried, creating the dark plains visible today. No wonder the ancients thought the lunar lowlands were Earth-like seas.

M. Carroll, *Picture This!: Grasping the Dimensions of Time and Space*, DOI 10.1007/978-3-319-24907-0_8, © Springer International Publishing Switzerland 2016

In the nineteenth and twentieth century, astronomers gained a much more accurate understanding of the lunar surface, and their knowledge informed the artists. Many depicted the mountains of the Moon as craggy spires. This was natural, as shadows along the terminator – the day/night edge – seemed long and pointed. The nineteenth century astronomer/meteorologist Abbe Theophile Moreux depicted lyrical, star-studded scenes of lunar craters beyond rugged Moon summits. Artist Paul Fouche illustrated astronomer Camille Flammarion's books, and included ragged mountain spires among his lunar depictions as well. Astronomer James Nasmyth's 1874 book *The Moon* took the view a step further. The author and his colleague James Carpenter constructed a series of plaster models based on their telescopic observations of lunar mountains and craters. Using harsh light, they photographed the models to illustrate their popular book.

However, the Moon's spikey shadows turned out to be an optical illusion. The pointed lunar shadows were simply elongated projections of peaks that, in reality, had a much more subdued and rounded profile. One astronomer knew this well. His name was Lucien Rudaux.

The quintessential grandfather of space art, Rudaux portrayed the Moon as a rocky, desolate world. Rudaux was director of the Meudon Observatory in Paris, France. Known as a keen observer, he applied what he saw in the eyepiece to canvas, as he was also an accomplished artist. Rudaux's careful observations of the lunar highlands illuminated against the sky, at the very edge of the Moon (the limb), led him to the correct conclusion that the lunar summits were quite rounded with gentle slopes.

Fig. 8.1 James Nasmyth's mountains of clay provided convincing lunar vistas to nineteenth-century readers (Image reprinted with permission from Amherst College Archives & Special Collections.)

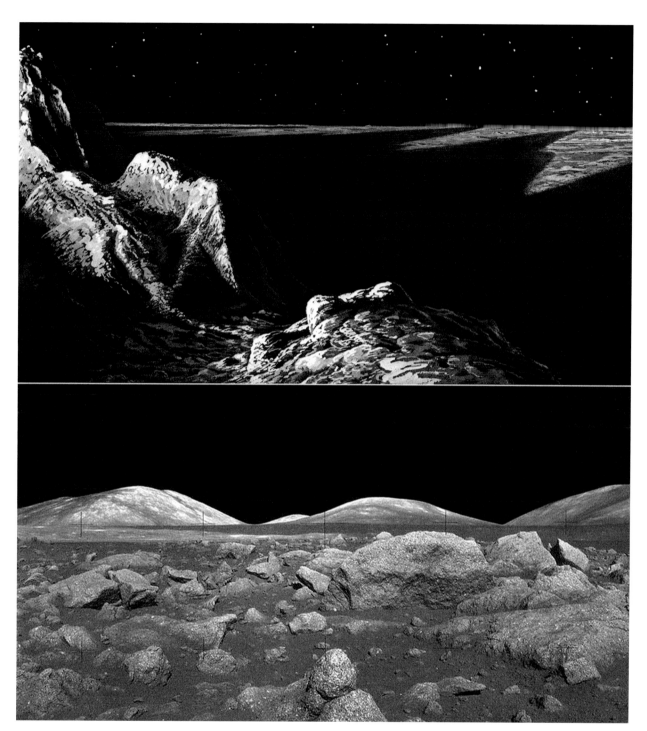

Fig. 8.2 Rudaux's impression of the lunar mountains, above, compares favorably with a photo of Camelot Crater taken by the crew of Apollo 17 (Estate of Lucien Rudaux, used with permission; Apollo 17 image courtesy of NASA/JSC.)

Rudaux's magnum opus was the landmark book *Sur les Autre Mondes* (*On the Other Worlds*).[1] Over 400 of his black and white and full color illustrations surveyed the lunar surface, the rocky deserts of Venus and Saturn's rings seen from within the planet's clouds.

[1] *Sur les Autres Mondes*, Larousse publishers, France, 1937.

Fig. 8.3 A carbonated ocean, painted by artist Paul Calle in 1966, seemed to be one of the reasonable possibilities to scientists, until Mariner 2 took the planet's temperature. At right, the reality seen through the eyes of the Soviet lander Venera 13 (Left: © Paul Calle, courtesy of Chris Calle; right: Venera image courtesy IKI, reprocessed by Don Mitchell and Michael Carroll.)

Venus provided lots of fuel for imaginative astronomers. The problem was not its distance (after all, it's the closest planet around), but clouds. A dense, opaque cloud layer covers the entire face of the planet, impenetrable to Earth's telescopes.[2] As Venus is also the most similar to Earth in distance to the Sun and physical size, it seemed reasonable that the neighboring world possessed fairly Earth-like conditions. But in 1932, spectroscopic studies revealed massive amounts of carbon dioxide in the atmosphere. Where was all that carbon dioxide coming from? Possible explanations of the Venusian climate ranged from it having a carboniferous swamp to its having deserts. Did vast pools of bubbling hydrocarbons blanket the surface, infusing the skies with vaporous carbon? If so, the clouds might consist of droplets of petroleum enough to make any Arabian oil baron jealous. Others envisioned a globe-swathing carbonated sea.

But to artists and scientists alike, Mars beckoned. The unclouded view of its changing surface made it a favorite among observers and poets. Ahead of any of his contemporary scientific illustrators, Rudaux excelled at his renderings of the Red Planet. While others depicted the world's dark smudges as swirling oceans, Rudaux swapped marine landscapes for desert ones. He rightly supposed that Mars was a dry world. Through the telescope, he and others had seen great dust storms move across the face of Martian plains. His Martian skies were even tinted with red dust, a

[2] That is, opaque to Earth telescopes of the visual-light type. The Arecibo radio telescope bounced radar off the Venusian surface, making the first crude radar maps of the surface in the 1970s.

phenomenon not confirmed until the *Viking* landers returned photos of the landscape in 1976. Rudaux's Martian vistas hold up well today, even after we have scoured the Martian plains and highlands with our orbiters, landers and rovers.

Of course, Rudaux was not the first to try his hand at Martian landscapes. Earlier panoramas reflected our warped view of Earth's next-door neighbor. That distorted interpretation came at the hands of a confluence of factors. First, Mars is a very difficult object to see. Although it is the second closest planet to Earth, it is small. Early observers had to strain at the eyepiece to discern the markings on its surface, and as they did so, they were able to figure out a few aspects of Mars's nature. Astronomers such as E. M. Antoniadi, Giovanni Schiaparelli, and a host of others realized early on that Mars had some significant Earthlike qualities. Its day times out at about 24 h and 40 min, and Mars has a season-producing axial tilt similar to Earth's. The polar ices ebb and flow with the seasons, as do ours. Additionally, the dark regions on the planet's face appeared slightly greenish, an illusion from telescope optics and relative color relationships. As seasons came and went, a "wave of darkening" visited the dark zones, as if vast forests were coming into bloom. This evidence led observers to believe they might be studying a planet very similar to Earth.

The planet seemed tangled in a net of straight and intersecting lines that also darkened with the seasons. The day's leading astronomers proposed that these lines might be channels carrying water, and wealthy Bostonian gentleman-astronomer Percival Lowell led the brigade ascribing an artificial intelligence to those "canals." The idea captivated the public and the scientific community alike, influencing music, literature, and the study of the Red Planet itself.

Lowell's canals continued to hold sway through the 1950s. Typical of the time is an illustration of a Martian waterway by Lowell Hess. The painting appeared in the children's book *Exploring Mars*[3] by Roy A. Gallant. By the decade of the 50s, evidence began to point to a thin Martian atmosphere. Nevertheless, many thought it likely that Mars's canals were real, perhaps natural phenomena related to geological fault lines or canyons. Most depictions of such natural formations included green areas near the liquid water. When *Mariner IV* flew by the Red Planet in 1964, the scientific community was nearly as shocked as the general public. The long-awaited smooth, canal-streaked plains did not show up. Instead, the first images boasted a cratered landscape more Moon than Mediterranean. As luck would have it, the first three flybys of Mars covered the most ancient, cratered terrain, just missing its sprawling canyons, towering mountains, and dry river valleys. Soon, a host of international orbiters transformed our view of Mars again, from dead back to the living, a world not Earthlike as much as uniquely Mars.

[3] *Exploring Mars* by Roy A. Gallant, illustrated by Lowell Hess, ©1956 Garden City Books.

Fig. 8.4 Lowell Hess's rendering of a Martian canal, ca. 1956. Hints of green vegetation rim the canal's edge (From Exploring Mars *by Roy A. Gallant, ©1956 Garden City Books.)*

By the time the Viking probes made landfall on the Martian plains, scientists knew that Mars's atmosphere was as thin as Earth's at 30,000 m. British master space artist David A. Hardy depicted the Martian sky in 1972, based on this data. His painting of astronaut explorers showed a deep blue sky fading into darkness, as was the expectation. The first *Viking* images from the surface changed all that. The Martian firmament glows with the tawny light of suspended dust particles, a surprisingly reddish hue. Hardy had to update his painting in 1978, airbrushing in a pale orange sky.

Over time, our view of the outer planets and their moons has morphed as much as has our conception of the terrestrials. In the past, the realm of the gas and ice giants seemed impossibly alien, with its numbing cold temperatures, vast distances and titanic orbs. The first renderings to show Jovian worlds beheld viewpoints from the supposed surface. Little did those artists and observers know that there was, in fact, no surface from which to watch those billowing ammonia storms. Early paintings of Jupiter's moon Europa included palm trees, and paintings of Jupiter itself displayed a world of ammonia-ice mountains and glowing volcanoes. Saturn's rings were seen as great flat hoops surrounding the golden planet, while Uranus and Neptune went largely ignored due to a

Fig. 8.5 *Astronauts explore the northern polar ices of Mars in this 1972 painting, which reflects the scientific thinking of the time. With the reality of a red Martian sky revealed in lander images, Hardy amended his image in 1978 (Paintings from* Challenge of the Stars, *© David A. Hardy. Used with permission.)*

Fig. 8.6 *The remarkable paintings of Ludek Pesek ranged from depictions of Martian dust storms (left) to views of Saturn from Mimas to a vista of Neptune (Paintings © Ludek Pesek, courtesy of Olga Shonova, Bruno Stanek. Used with permission.)*

lack of information. But not by all. Czeck artist Ludek Pesek painted a remarkable collection of planetary landscapes in the 1960s and 1970s, including views of the chiefly overlooked Uranus and Neptune. His depictions of astronomical subjects earned him the honor of having an asteroid named after him, 6584 Ludekpesek.

Fig. 8.7 Astronomical artists had little to go on as they tried to project what Pluto might look like. Nevertheless, some guesses came close. In 1979, Don Dixon painted Pluto as seen over its moon Charon (above left), presciently close to the reality seen by New Horizons (upper right). David A. Hardy offered this view of Charon in 1999 (lower left), comparing favorably to the real moon at lower right (Paintings courtesy of Don Dixon and David A. Hardy. Used with permission.)

Today, these worlds of gas and ice intrigue us, as do their remarkable, palm tree-less moons. Humankind possesses an inquisitive streak. As soon as these worlds became real to us, we wanted to go. But how would we get there? The next chapter will answer that question.

Chapter 9
Space Travel

A few centuries ago, travel to the distant planets took the form of fanciful trips using swan-drawn carriages, gigantic cannons or even dreams. But with the advent of rocket propulsion,[1] voyages to the Moon and beyond took on a much more serious tone. Scientists and engineers began to thoughtfully plan strategies for getting people out into the "ether." How did we do? By comparing the visions of the past to the actuality that unfolded in our human space exploration programs, we can explore the scale of time – and the progress of our understanding of the universe – in concrete ways.

Early on, the most obvious target of exploration in the sky was the closest and most carefully observed – our own Moon. It sat there, up in the night sky, taunting us, daring us to visit its mysterious dark seas. With the introduction of rocket technology, engineers gained a grasp of how we might actually get there. A rocket engine works in a vacuum, and as scientists realized that a vacuum or near vacuum existed beyond Earth's atmosphere, designers rose to the task of voyaging through it.

Serious rocket propulsion engineering came at a remarkably early stage, as did the associated designs needed to enable an astronaut to survive such a voyage. Founded in Liverpool in 1933, the British Interplanetary Society's stated goal was to promote astronautics and human space exploration. From 1938 to 1947, the society carried out one of the first serious feasibility studies for a human expedition to the Moon. The BIS envisioned a two-story landing craft with shock-absorbing legs. The lower stage would remain on the surface, serving as a launch pad for the upper portion of the craft, the Earth return vehicle. The prescient British design called for a crew of three. Explorers were to remain on the lunar surface for two full weeks (half a lunar day). The craft would return home with half a ton of lunar treasures. In fact, this was exactly the strategy, in modified form, used by both the American and Soviet lunar programs. The concept came to be known as the BIS lunar spaceship, and, in the words of the BIS, "must be regarded as one of the classical pioneering studies in the history of astronautics."

The detailed engineering of the BIS Moon lander far surpassed anything that had come before it and showed clever solutions to various problems of space travel. The crew cabin was made of plastic, and was protected during liftoff by a ceramic cupola. The crew rode into space in form-fitting couches made of rubberized horsehair fabric. A catwalk encircled the wall of the cabin and would serve as a floor as the ship spun in transit. Four windows faced forward, with another 12 embedded in the walls. Six more were inserted into the floor. The crew could also make use of "coelostats," instruments that enabled them to constantly monitor the

[1] The Chinese made use of solid rockets as far back as the eleventh century. Records from 1264 mention the "ground rat," a kind of firework. But it was not until the early twentieth century that engineers began to study rockets as a mode of space propulsion. Tsiolkovsky's *The Exploration of Cosmic Space by Means of Reaction Devices* advocated the use of liquid oxygen and hydrogen to fuel rockets.

M. Carroll, *Picture This!: Grasping the Dimensions of Time and Space*,
DOI 10.1007/978-3-319-24907-0_9, © Springer International Publishing Switzerland 2016

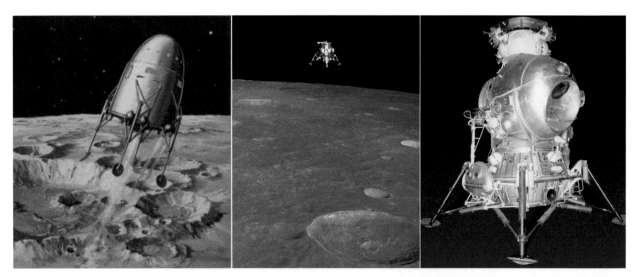

Fig. 9.1 The British Interplanetary Society's plans for a lunar landing vehicle, illustrated by R. A. Smith, bear a striking resemblance to NASA's lunar lander (center), and the Soviet LK lunar lander (right). In common were such features as folding landing legs, ladders for egress, and a two-stage approach using the lower stage as a launch platform for the return (Left: Image courtesy of the British Interplanetary Society. Center: Image courtesy of NASA. Right: Image © Michel Koivisto, courtesy Karl Dodenhoff, myspacemuseum. com. Used with permission.)

stars for navigation, even while the ship rotated for gravity. Artist/designer Ralph A. Smith, a member of the society, rendered a series of beautifully detailed illustrations of the Moonship while in flight and during exploration on the lunar surface.

As designs matured throughout the post-World War II era, other studies began to show up in children's books, many influenced by the BIS study. Mae and Ira Freeman penned the story *You Will Go To The Moon*,[2] an optimistic voyage from Earth to orbiting space stations, and then on to Moon bases. While couched in a children's literature format, the book serves as a fine overview of our general concepts at the end of the 1950s.

At about the same time, Golden Books released *The Moon, Our Neighboring World* by Otto Binder. The text describes a lunar outpost:

> *Scientists speak of cities nestled within craters, covered by plastic domes and comfortably air-conditioned. The moon's low gravity-pull (1/6 of earth's) will allow the men to travel by leaping 100 feet at a bound, like huge frogs…Metals would not tarnish on the moon for centuries as there would be no oxygen or water to rust them. Fresh foods would keep indefinitely without spoiling since there are no living bacteria on the moon to cause decay… Rocket ships could take off and land much more easily in the moon's low field of gravity….*

At the end of the 1950s, U. S. rocket designer Werner Von Braun teamed with artist Fred Freeman to craft a serialized adventure to the Moon, appearing in *This Week* magazine, and then as the book *First Men to the Moon*.[3] What the story lacked in character and plot development, it made up for in meticulous engineering detail.

[2] *You Will Go To The Moon* by Mae and Ira Freeman, 1959, Beginner books/Random House.

[3] *First Men to the Moon* by Werner Von Braun, 1958, Holt, Reinhart & Winston.

Fig. 9.2 Otto Binder's caption for George Solonovich's illustration said, in part, "Reaching the moon by spaceships, future explorers will find a strange bleak world…" Note the uncovered fuel tanks, common to early designs (The Moon, Our Neighboring World, by Otto Binder. © 1959 by Golden Press, Inc.; Public domain)

Fig. 9.3 Left: *"You could not climb this kind of hill at home. But you can here!" Caption for this illustration from* You Will Go to the Moon *(Beginner Books, Random House 1959). Note the similarity of the craft in the background to the BIS design.* Right: *"Firing Off the Seismograph Rockets" by Fred Freeman, from Von Braun's book* First Men to the Moon *(Image courtesy of Frederic W. Freeman Trust. Used with permission.)*

Von Braun described the kind of suit needed for the lunar surface. Before *First Men to the Moon*, futurists imagined space suits similar to deep-sea diving outfits, with heavy, bolted helmets and rigid, claw-like gloves. But Von Braun's sleek suit featured a tinted helmet visor, a radio antenna on a backpack, pressure tanks, and wrist-mounted gauges, all features of the real

Apollo Moon suits. Von Braun's air tanks carried an oxygen/helium mixture. Another prescient idea in the book involved the use of explosive rockets to aid in seismic studies. The later Apollo flights carried seismic explosives. And in another prophetic portrayal, his astronauts even used a "descent ladder" to exit their landing craft, just as the "dusty dozen" of the Apollo era did.

Von Braun also outlined various trajectories that could be flown by astronauts en route to and from the Moon. His astronauts made a mid-course correction before entering Earth's atmosphere, a routine maneuver by Apollo crews. His ship landed like an aircraft, quite reminiscent of the later space shuttles.

Von Braun's engineering work led to the most powerful booster ever flown, the Saturn V. The Saturn and other boosters rode the shoulders of thousands of years of experimentation. The first true rockets accompanied the Chinese invention of gunpowder. In celebrations, first-century partiers filled bamboo tubes with gunpowder, setting off the sparks in celebration. By the thirteenth century, these tubes had evolved into glorified bottle rockets, and took to the skies as weapons of war in a Chinese/Mongol battle.[4] Three hundred years later, German fireworks designer Johann Schmidlap constructed a "step-rocket," a multi-stage craft that delivered his fireworks to higher altitude. Modern rocketry was borne in the mind of a Russian school teacher, Konstantin Tsiolkovsky, who realized that a rocket engine could operate in a vacuum, pushed along by the force of escaping gas.

Robert Goddard built the first liquid-fueled rocket in 1926. From that point on, rocket propulsion technology took off. German V2 rockets embodied the best of the engineering in the 1940s. Efforts to explore space culminated in a progression of bigger and better boosters: Atlas, Titan, Saturn V, Proton, Energia, Ariane, Falcon 9, the space shuttle and others. From the first Chinese fireworks to the mighty Saturns, rocket power grew from a few pounds of thrust to Saturn's 34,020 kN (see Fig. 9.4).

Some predictions of space travel were hauntingly accurate. A typical series of examples graced the pages of a novel written by the astronomy artist Ludek Pesek.

THE MOON OF LUDEK PESEK

[4]A fascinating Chinese legend describes what may have been the first rocket-powered human flight. A city official named Wan-Hu supervised construction of a rocket-propelled flying chair. Two kites were attached to the chair top, with 47 rockets affixed to its back. While his launch must have been spectacular, his landing was not documented.

In 1964, 2 years before *Luna 9* and *Surveyor 1* gave us our first human-scale views of the Moon from its surface, Czeck master space artist Ludek Pesek wrote a "children's" book called *Log of a Moon Expedition*. His narrative detailed the first expedition to the Moon. Its science was as sophisticated as the best mainstream science of the time, and informed the action as well as the art.

Pesek did a series of eight paintings for the book, all in black and white. He got a few things wrong, and a lot right. It is entertaining and informing to compare the science fiction of *Log of a Moon Expedition* to the reality that came just a few years later. Here are a few examples.

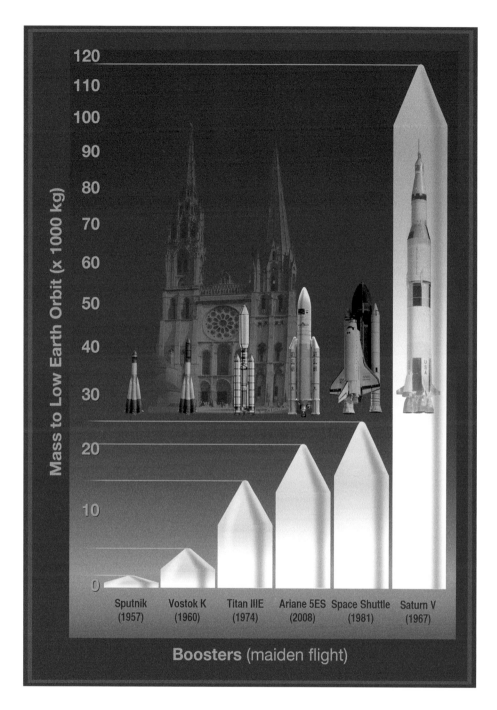

Fig. 9.4 A comparison of power. The muscle of rocket boosters has increased throughout history, beginning with the tiny Chinese "fire arrows" (too small to show) and culminating in the powerful Saturn V. The great cathedral of Chartres is shown behind the launch vehicles, to scale (Chartres photo and art by the author.)

Humans have now walked across the dusty lunar landscape. Twelve brave souls visited that high frontier from 1969 to 1972. Many scientists and space planners see great value in returning there, adding to what some see as an initial survey needing modern follow-on exploration. Others have set their sights higher. Some space experts advocate Mars as the next logical step in humanity's ventures into the cosmos.

Fig. 9.5 *Pesek's first lunar explorers brought two rovers with them. One of his paintings is strikingly familiar to fans of* Apollo 15, *which is seen here in front of the 4.6-km-high Mount Hadley (Pesek painting courtesy of Olga Shonova, used with permission; Apollo NASA photo.)*

Fig. 9.6 *Pesek describes the lunar surface as having practically no dust. His heroic crew must contend with many fractures and chasms in their rocky journey. At right: Apollo 15's* Jim Irwin *at Hadley Rille (Pesek painting courtesy of Olga Shonova, used with permission; Apollo NASA photo.)*

Fig. 9.7 *Pesek's story has the crew of eight erecting several communications antennae. Similarly, the* Apollo 12 *crew deployed such an antenna for transmission of color TV. Sadly, the camera was damaged when it was inadvertently pointed at the Sun (Pesek paintings courtesy of Olga Shonova, used with permission; Apollo NASA photo.)*

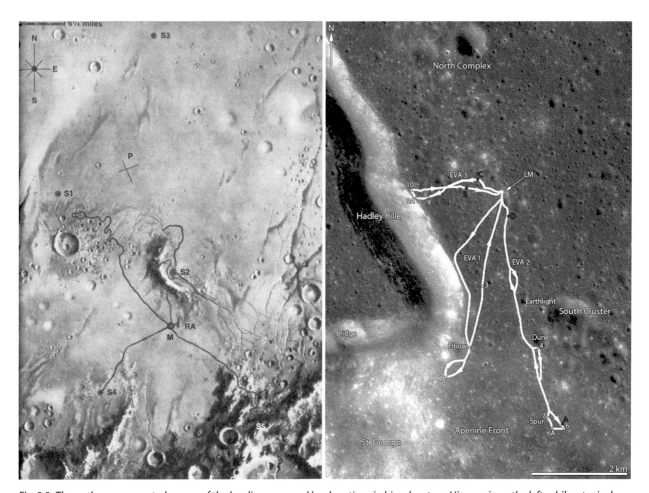

Fig. 9.8 *The author even created a map of the landing area and key locations in his adventure. His map is on the left, while a typical Apollo map, this one from* Apollo 15, *is on the right (Pesek's book* Log of a Moon Expedition *is still available through used and rare booksellers. The first American edition, released in 1969, was through Alfred A. Knopf in Canada by Random House, and originally published in Germany by George Bitter Verlag.) (Pesek map courtesy of Olga Shonova, used with permission; Apollo map NASA.)*

MARS

We do not need to travel as far back in time to see studies for Mars exploration. Wernher Von Braun and Willey Lay laid out a carefully crafted blueprint for an ambitious Mars expedition, first conceived in 1948. Von Braun's plan was an ambitious one, initially involving an armada of ten huge spacecraft carrying a total crew of 70. Once in orbit around Mars, von Braun's plan called for three winged rocketplanes to descend to the surface. The first of the three landers would set down on the polar ice cap, using skids instead of wheels. Von Braun reasoned that the polar ices would be a smooth place for a safe landing site. The huge craft would carry tread-equipped rovers and provisions for the other explorers, who would come down later. First, the surface crew had to drive 4000 miles to the Martian equator. There, they would use their tractor to fashion a landing strip for the other two planes. Von Braun's massive expedition would remain on Mars for 15 months, waiting for the next launch window to return home.[5] Removing the wings from their rocket plane, crews would pivot the central portion of the planes vertically for takeoff, an event probably yearned for after 15 months in the Martian wilderness.

By 1956, the scale of Von Braun's mission seemed overly ambitious. He streamlined his mission architecture to two ships and 12 crewmates. The plan unfolded before the public as a series of articles for *Colliers Magazine*, later becoming the book *Exploration of Mars*, which he coauthored with Willey Ley.[6] The book's true magic came in the form of its spectacular illustrations by master space artist Chesley Bonestell.

[5] Launch windows, or opportunities, occur roughly every 18 months between Earth and Mars.

[6] *Exploration of Mars* by Werner Von Braun and Willey Ley, Viking Books, 1956.

Fig. 9.9 This Chesley Bonestell masterpiece shows Von Braun's Mars flotilla in Earth orbit, preparing for departure to the Red Planet (Image © Chesley Bonestell, courtesy of Bonestell LLC. Used with permission.)

Fig. 9.10 The Mars Excursion Module, part of a 1963/1969 NASA study to send humans to Mars. The bizarre shape of the craft reflected "lifting body" aerodynamics, which incorporated the shape of the craft itself to aid in lift as it descended through the atmosphere. The space shuttles used lifting body profiles. The 1989 NASA Lewis Research Center study of a Mars outpost, painting by Les Bossinas (Both photos courtesy of NASA.)

Von Braun was not the last to fashion an overly ambitious Mars plan. With the successes of the Apollo program, NASA drew up blueprints for follow-on expeditions to the Red Planet. These usually took the form of Apollo on steroids, with bigger rockets and far bigger budgets. During the Nixon administration, Vice President Spiro Agnew tasked NASA with taking what many saw to be the next logical step, a human mission to Mars. Agnew's timing was poor: the United States was enmeshed in the Vietnam War, and recession plagued the economy. Congress shelved the plans as too expensive.

That didn't stop designers from further exercises in excess. During the George H. W. Bush administration, NASA once again floated a huge Mars mission, which the press referred to as Battlestar Galactica (after a popular science fiction television show of the time). The mission had high goals and higher price tags, using the same sky-is-the-limit scales as had been proposed in the past. One mission would have cost more than the entire Apollo program. The plan never got off the drawing board.

Modern plans for the exploration of Mars cover a wide range of scales, but usually involve comparatively small crews of three to five, with multiple launch vehicles. NASA is developing the capabilities needed to send humans to an asteroid by 2025 and Mars in the 2030s. NASA plans presented in 2010 call for a robotic mission to capture and return an asteroid to lunar orbit. Astronauts would travel to the Moon's environs and explore the asteroid some time in the 2020s, returning samples to Earth. NASA would apply experience gained from the mission to a future Mars expedition. The plan depends on an advanced version of NASA's next generation Space Launch System, a booster larger than the Saturn V. But the SLS is huge and expensive, so other designers are looking for more inexpensive mission scenarios.

On the forefront of this effort is the Mars Society's Robert Zubrin. Zubrin's Mars plan is streamlined, with much redundancy built in. Involving fewer vehicles and far less expense, his mission architecture depends on the manufacture of fuel on the Martian surface. Using technologically simple gas manufacturing technology, Zubrin's mission would begin by landing an unpiloted craft on the Martian surface. The craft would later serve as a launch vehicle to return astronauts from the Martian surface to a ship waiting in orbit. The trick – the vehicle would land on Mars with empty fuel tanks, making it lighter and requiring less expense in the launch from Earth. While on the surface, the craft would manufacture its own fuel using the Martian atmosphere. Its propellant factory could combine 1 ton of hydrogen from Earth with elements of carbon and oxygen from the Martian atmosphere, producing 20 tons of fuel for the return trip. When the craft radioed that it was full, a crew could be safely dispatched to Mars, carrying only enough fuel to get them there. But at the same launch window used by the crew, flight engineers would send a second ascent vehicle, which would land when the crew did and immediately begin to synthesize its own fuel. By the time the crew was ready to return home, two return vehicles would be fully fueled and available for the return trip. In the meantime, multiple other spacecraft would be in orbit, providing a safety margin for all stages of the flight to and from Mars. Zubrin's vehicles could all be launched on fairly modest boosters. One scenario utilizes SpaceX's Falcon 9 to launch the various elements. Falcon 9 is the largest privately designed booster, fabricated and flown by Elon Musk's SpaceX company. The craft has successfully supplied the International Space Station several times, and plans move forward for expanded development of SpaceX's fleet.

Utilizing Zubrin's creative approach, Mars Direct weighs in at a total mission mass of about 88 tons. Past NASA scenarios incorporated equipment and vehicles totaling over 600 tons. Mars Direct comes in at an estimated price tag of $30 million, compared to the earlier NASA Battlestars of $450 million. But NASA has paid attention, and their most recent scenarios echo some architectures from Zubrin's designs, albeit in a subdued fashion. In-situ fuel production, creative manufacturing and testing of vehicles, and other cost-cutting strategies are showing up in the more recent NASA plans. The multiple vehicles, redundancy, and in situ resource utilization of Mars Direct offers something no other plan does: its goal is to establish a permanent pathway between Earth and Mars.

Across the pond, Russian aerospace giant RKK Energia is undertaking studies outlining techniques to build a human Mars mission based on the International Space Station's Zvezda module. Their plans envision the 77-ton craft carrying a crew of six to Mars. Costing upwards of $15 billion, depending on international contributions, the proposed mission would last 900 days. Designs call for solar electric propulsion supported by solar panels that would span seven times the length of a football field. The new generation of Mars travel blueprints by NASA, Mars Direct and Energia provide a contrast to the visions of 70-member crews and rocketplanes that came before.

FARTHER TRAVELS

As for projections beyond Mars, traveling down our time scale into the distant past is unnecessary. Although science fiction authors have been exploring such trips for nearly a century, only in the past few decades have engineers seriously explored sending humans to the outer Solar System. Scenarios often involve artificial gravity for long flights, hibernation, advanced propulsion to reduce trip times, and the generating of magnetic fields for radiation protection.

By the Apollo era, other visionaries had begun to seriously consider the technologies necessary for outer planets exploration and settlement. In his book *Challenge of the Stars*,[7] British artist David A. Hardy teamed with astronomer Patrick Moore to describe (and paint a picture of) an outpost on Jupiter's inner moon Amalthea. In part, the authors' caption read, "The observatory is designed chiefly for studies of Jupiter itself; to the right can be seen the laboratories, formed from the cylindrical tank-sections of the propulsion module which brought the expedition to Amalthea."

Hardy and Moore go on to describe mining operations on Saturn's moon Titan, and voyages even further out. Their narrative culminates in projections about Pluto exploration. They say, "Long before manned craft can venture out into these depths, we ought to have close-range photographs of Pluto…After having bypassed Pluto, the probe should still maintain radio contact with Earth." At the time, no specific plans existed for such an interplanetary craft, but the New Horizons Pluto probe fits the book's description perfectly.

With the completion of the Voyager flybys of Jupiter and the other giants, projections became more focused and realistic. In their masterwork *Out of the Cradle*,[8] William K. Hartmann, Ron Miller and Pamela Lee offered a richly illustrated smorgasbord of planetary exploration. In their chapter "Into the Realm of Ice and Fire," the author/illustrators sent their astronauts touring territory from the Galilean satellites of Jupiter to the co-orbital moons of Saturn, with panoramas of ice giants and Pluto dropped in for good measure. Concentrating on the Jupiter system, readers feasted their eyes on astronauts landing on the smooth surface of Europa, soaring over Callisto's Valhalla impact basin, and gazing at a massive Jupiter from a cave on Io. But *Out of the Cradle* preceded the better data culled from missions such as the Galileo orbiter at Jupiter, which showed us just how toxic the Jovian environment is to human explorers. While astronauts take a leisurely stroll on the moon Io in the book, the reality is that Jupiter's radiation would bathe astronauts in radiation ten times the lethal dose in the span of a few hours. Even further out at Europa, daily radiation levels equal twice that of the human lifetime allotment. Advanced concepts include ships equipped with radiation "storm shelters," but the complexities of travel in the outer Solar System are varied and plentiful.

[7] *Challenge of the Stars* by Patrick Moore and David A. Hardy, Mitchell Beazley ltd/Random House, 1972.

[8] *Out of the Cradle*, William K. Hartmann; ©1984 Workman Publishing Company, Inc.

Fig. 9.11 *This 1972 painting by David A. Hardy shows an outpost on Jupiter's moon Amalthea. The moon was not imaged in detail until the Voyager encounters 7 years later (Image © David A. Hardy. Used with permission.)*

The depiction of outer planet exploration came to maturity in Sebastian Cordero's stark film *Europa Report*. The movie portrays a privately funded venture to search for life on Jupiter's oceanic ice moon. Private sector involvement may be a game-changer in near-future human exploration of the Solar System. It has already made its impact on the servicing of the International Space Station, for example with SpaceX's Dragon capsules.

Fig. 9.12 From the book Out of the Cradle, *Pamela Lee depicts astronauts in one of the steep-walled grooves of Callisto (top). The caption reads, in part, "Here, astronauts …cut into a groove wall to obtain a cross section of structure of the ice/soil mixtures, in hopes of clarifying the history and composition of the region." In the lower image, Lee shows us the view explorers would behold from a cave on Io (Paintings ©1984 Pamela Lee. Used with permission; From the book "Out of the Cradle" by William K. Hartmann, Ron Miller and Pamela Lee, Workmann, 1984.)*

As for engineering, *Europa Report's* ship featured a spinning section arranged as a rotating dumbbell behind the stationary crew cabin. The main cabin also served as the descent vehicle. Astronauts carefully monitored radiation levels, something not on the radar screen of earlier Jovian tales.

One day, when humans finally do venture across the icy plains of the Galilean satellites, *Europa Report* will undoubtedly entertain future audiences with its "parochial" imaginings. But before human exploration becomes reality, the von Brauns, Peseks and Corderos must pave the way.

Our travels have taken us from the smallest of asteroids to the largest of galactic structures, and from warped time around gravitational lenses to the "deep time travel" of light coming from the universe's very beginning in both space and time. Whether we talk about the size of objects, the scale of planetary or star systems, or the timeline along which we live, scale provides a critical tool and framework for our understanding of the universe around us.

Index

M. Carroll, *Picture This!: Grasping the Dimensions of Time and Space,*
DOI 10.1007/978-3-319-24907-0, © Springer International Publishing Switzerland 2016